Textbook of
GYMNOSPERMS

The Authors

Dr Inderdeep Kaur is an Associate Professor in Botany, Sri Guru Tegh Bahadur Khalsa College, University of Delhi. She has an experience of teaching Gymnosperms to the undergrads for over fifteen years. With papers of repute in international and national journals, she has been a referee for various research papers. She has also been a participant in various conferences abroad and in India. Dr Kaur has authored chapters on various topics in Biology, has co-edited a book entitled "Plant Reproductive Biology and Conservation" and co-authored a book "Brown Algae: Structure, Ultrastructure and Reproduction". She has been Principal Investigator of projects under Department of Science and Technology and University Grants Commission.

Prof. (Dr) Prem Lal Uniyal is a Professor of Botany, University of Delhi. With specialization in Archegoniates, he has been engaged in teaching post graduate level for over 25 years. He has published nearly 100 research papers, two books and thirty chapters and has also guided students for PhD. His current area of interest is conservation of Archegoniates.

Textbook of
GYMNOSPERMS

Based on CBCS Syllabus of University of Delhi

Inderdeep Kaur

Associate Professor in Botany
Sri Guru Tegh Bahadur Khalsa College
University of Delhi, Delhi

Prem Lal Uniyal

Professor in Botany
University of Delhi, Delhi

Daya Publishing House®

A Division of

Astral International Pvt. Ltd.

New Delhi – 110 002

ISBN 978-93-5919-575-9

Published by : **Daya Publishing House®**
A Division of
Astral International Pvt. Ltd.
– ISO 9001:2008 Certified Company –
4760-61/23, Ansari Road, Darya Ganj
New Delhi-110 002
Ph. 011-43549197, 23278134
E-mail: info@astralint.com
Website: www.astralint.com

Printed at : **Replika Press Pvt. Ltd.**

Preface

Gymnosperms is a widely distributed and commonly encountered group of plants. These plants are found in cooler areas and have great commercial value. By evidence from morphological, molecular, phylogenetic and evolutionary developmental studies, considered as a sister group of angiosperms. The plants which dominated vegetation in Mesozoic era - gymnosperms are now a threatened group.

The previous classifications proposed for gymnosperms were based on characters which were rather ill defined. New trends in evolution biology focus on gymnosperms as an important link in the evolution of flowering plants. Understanding of inter-relationships within various groups has therefore, become important. Though the phylogenetic studies have remained inconclusive, recent linear classification by Christenhusz *et al.* (2011), followed in this book has tried to resolve the issues to a large extent.

The book has seven chapters in all besides colour plates of five genera - *Cycas, Ginkgo, Gnetum, Ephedra* and *Pinus.* The first chapter is an attempt to introduce GYMNOSPERMS with some of the lesser known facts *e.g.*, the hydrophobic nature and reasons for the present threatened status of the group. Four subclasses with representative genera are discussed in detail in chapters 2-5. The format of presentation introduces the subclass followed by a detailed morphological, anatomical and reproductive account of the specific genus. Since each subclass is distinct and has special characters to offer, these updates are given in boxes or have been highlighted in the text. Adding a few questions at the end of each chapter would allow students to check their understanding and grasp on the specific study. Special attention has been paid to illustrations that convey a much clearer conception of structural features of stages in reproduction. Besides black and white diagrams/

photographs, colour plates have been provided at the end of the book which are planned in context with the laboratory work.

The last two chapters (6 and 7) are added to broaden the scope of the book and make it of value to students in general, and those interested in research or knowing about the utility of these plants. We have also used a consistent terminology for the reproductive structures. Term cone is used only for Pinidae where the reproductive structures are compact and woody ,for greater part of reproductive phase while in rest of the subclasses, term strobilus is used because the reproductive structures are loose and also not woody.

We have also made an attempt to provide information in a lucid style which builds the chapter point wise and expect it to make the study easy and interesting for the students. The question bank provided as part of the appendix is a collection of questions that have appeared in Examinations of University of Delhi and will be of immense help to students preparing for competitive examinations.

The text is primarily designed for the undergrads pursuing both Honours and Life Sciences courses and also for students who opt for Generic Elective papers on Biodiversity. A long association with students while teaching papers on ARCHEGONIATES, made us realize the problems they encounter while learning about gymnosperms, and preparing for examinations. The content has therefore, been presented in a very concise and easy- to- remember format. The information is up to date and recent trends in our understanding of first land plants with seeds are provided. The book also provides a detailed account of topics like pollination drop, double fertilization, post pollination and post fertilization events in various genera. The colour plates will be appealing and helpful while handling practical sessions. This book will cater to the needs of both learners as well as teachers.

We would like to express our great appreciation to our students who motivated us into authoring this book. The thoughtful and insightful interactions with them during theory and practical classes helped us shape the vision of the book and profoundly enriched the contents.

We are indebted to Dr Surinder Kaur, Associate Professor in Botany, Sri Guru Tegh Bahadur Khalsa College, University of Delhi for her invaluable advice and perspective criticism through all stages of the preparation of the book.

We also acknowledge the stimulating discussions on several topics with Dr. Anjana Sagar, Assistant Professor in Botany, Sri Guru Tegh Bahadur Khalsa College, University of Delhi.

Special thanks to Prof. Rupam Kapoor, Department of Botany, University of Delhi, and Dr Monika Koul Moza, Assistant Professor in Botany, Hans Raj College, University of Delhi, for their valuable support and continued inspiration.

Thanks are also due to Ms. Garima Anand and Ms. Priti Giri Ph.D. scholars, Department of Botany, University of Delhi for their support throughout book writing.

We are especially indebted to Dr Jaswinder Singh, Principal, Sri Guru Tegh Bahadur Khalsa College, University of Delhi, Delhi and Prof. K. S. Rao, Head, Department of Botany, University of Delhi, Delhi for encouraging our scholarly endeavours.

Thanks to our publisher Daya Publishing House, A Division of Astral International Pvt. Ltd for not only providing outstanding technical support but also for suggesting intricate modifications. Lastly, we also wish to thank Mr. Tanuj for designing the cover of the book.

Inderdeep Kaur

Prem Lal Uniyal

Contents

Preface v

1. Introduction 1

General Features – Chronology of Gymnosperm Evolution – Are Gymnosperms Hydrophobes? – Gymnosperms vs Angiosperms – Classification

2. Subclass Cycadidae 20

Are cycads living fossils?
Cycas – Distribution – Sporophyte – Morphology – Anatomy – Reproduction – Gametophytes – Pollination – Fertilization – Embryogeny –Seed

3. Subclass Ginkgoidae 45

Is Ginkgo a living fossil?
Ginkgo – Distribution – Sporophyte – Morphology – Anatomy – Reproduction – Gametophytes – Pollination – Fertilization – Embryogeny – Seed

4. Subclass Gnetidae 61

Gymnosperms with obscure affinities.
Gnetum – Distribution – Sporophyte – Morphology – Anatomy – Reproduction – Gametophytes – Pollination – Fertilization – Embryogeny – Seed Relationships of Gnetales – Gnetum and Angiosperms – Economic Importance

Ephedra – Distribution – Sporophyte – Morphology –Anatomy – Reproduction – Gametophytes – Pollination – Fertilization – Embryogeny – Seed – Salient Features of Sexual Reproduction – Economic Importance

5. Subclass Pinidae **99**

The success story of conifers.
Pinus – Distribution – Sporophyte – Morphology – Anatomy – Reproduction – Gametophytes – Pollination – Fertilization – Embryogeny – Seed

6. Gymnosperms and Human Welfare **129**

As Ornamentals – As Food – As Medicines – As Wood – As Paper Pulp – As Paints and Varnish – As Oil Used in Scents – Miscellaneous uses

7. Collection and Processing **134**

Collection of Woody Plants – Preservation – Staining – Need for Digital Herbaria

COLOUR PLATES 138

Glossary 152

Question Bank 155

Suggested Readings 158

Index **163**

"C is for Conifers"

Most with cones for seeds, most with needles for leaves
C is for conifers, my kind of trees.
Large junipers and fir, spruce, cedars and pines
C is for conifers, more than five hundred kinds.
If you see a Christmas tree or a stack of newspapers or a 2x4 frame of a house
It's probably made from pine trees and pine trees are conifers, that's what this song is about.
Or if you see a plant in the shape of an elephant, or in the shape of a dog
It's probably a shrub, a conifer shrub pruned into that shape by someone.
Most with cones for seeds, most with needles for leaves
C is for conifers, my kind of trees.
Large junipers and fir, spruce, cedars and pines
C is for conifers, more than five hundred kinds.

Spoken: There are so many different kinds of conifer trees.
This is just a few of them: Spruce,
Hemlocks, larch, cyprus, juniper,
Fir, douglas fir, cedar, yew,
Pine, the dwarf conifer,
And the great California sequoia redwood.

Most with cones for seeds, most with needles for leaves
C is for conifers, my kind of trees.
Large junipers and fir, spruce, cedars and pines
C is for conifers, more than five hundred kinds.

Source: http://tmbw.net/wiki/Lyrics:C_Is_For_Conifers

Artist: They Might Be Giants

Album: Here Come the ABCs

Released: 2005

Genre: Children's music

Chapter 1

Introduction

Gymnosperms, the first seed plants evolved around late Devonian period. By the end of Paleozoic era, they became competitive enough, replacing mosses, ferns and horsetails, only to become the dominant vegetation of Mesozoic era. The era is therefore, also known as 'ERA OF GYMNOSPERM DOMINANCE'. Appearance of flowering plants, the angiosperms during the Cretaceous period posed a tough competition for gymnosperms, which then got restricted to certain habitats.

Gymnosperms are widely distributed, woody, seed-bearing plants but, with naked seeds. They are commonly encountered in arboreta, with conifers dominating particularly the north temperate mountains. The relationships among gymnosperms have been an open discussion for quite some time, with recent molecular work over the past two decades, having resolved taxonomic problems to a large extent.

The group differs from angiosperms in not having an ovule enclosed in a carpel. Instead, the fertilized ovules develop into seeds on the surface of integuments (interpreted as a scale, bract or leaf or sporophyll) that are aggregated Into cones or cone-like structures **(Plate V.5&8)**, with the exception of *Cycas* (Cycadaceae) **(Plate I.6&9)** and *Ginkgo* (Ginkgoaceae).

Gymnosperms are slow growing trees taking up to a year for pollinated ovules to be fertilized and some take up to another three years to mature. The reproductive structures are usually unisexual and organized into cones or cone-like structures also known as strobili.

The plants are currently divided into four groups: cycads (Cycadidae), ginkgos (Ginkgoidae), conifers (Pinidae) and three enigmatic genera in the Gnetidae. Morphologically these groups are very different from each other but all four groups have ovules unprotected prior to fertilization, a synapomorphy for gymnosperms.

Extant gymnosperm groups occupy diverse habitats and some have probably survived after making adaptive shifts. Gymnosperms are considered to be highly conservative in their morphology and have not changed much over the period of time. As a result, a number of species for example, *Metasequoia glyptostroboides*, *Ginkgo biloba* and *Wollemia nobilis* (Niklas, 1997), *Ephedra* (Rydin *et al.*, 2004) and cycads in general (Norstog and Nicholls, 1997) are referred to as 'living fossils' because they are morphologically very similar to Mesozoic fossils. Gymnosperms have much lower species diversity than angiosperms (c. 900–1200 vs 250 000–270 000 or more, respectively; Hill, 1998; Magallon and Castillo, 2009; Bell *et al.*, 2010; Christenhusz *et al.*, 2011; Earle, 2011), exemplified by many high-level taxa with one (Ginkgoideae, Sciadopityaceae, *Metasequoia, Wollemia*) or a few species, while no gymnosperm genus has more than 120 species. Today, there is a huge difference in species diversity between gymnosperms and angiosperms that has developed over approximately 350 my (million years) since their lineages diverged.

The biodiversity maps and analyses at continental to global scales on gymnosperms are scarce (Pimm *et al.*, 2014). Regarding gymnosperms, little synthetic work has been conducted at a global scale, so only broad conclusions about their pattern of diversity can be formulated. Moreover, geographical distributions of many gymnosperm taxonomic groups remain poorly documented (Kreft and Jetz, 2007).

Recently, several comprehensive compilations of gymnosperms have been published providing us with a broad evolutionary overview of the group.

Extant gymnosperms are not numerous. There are about 1026 species in all, though the figures in surveys conducted over a period of time vary a little perhaps due to incomplete documentation. The three 'non - conifer' groups comprise ca (approximately) 310 species of cycads in 10 genera, one extant ginkgophyte and 80–100 gnetophytes in three genera. According to Farjon (2010) there is a total of ca 615 species of conifers in 70 accepted genera **(Table 1.1).**

Table 1.1 A - C: Total number of genera and species under various groups as per surveys carried out.

A&B - Data is based on studies till 2002.07.07 (after Christopher – Gymnosperm database). C-The total number has been revised after 2010 survey (Also see Kozlowski *et al.*, 2015).

			Family	Genera	Species
A	Gymnosperms	Pinidae	Araucariaceae	3	40
			Cupressaceae	30	142
			Pinaceae	11	232
			Podocarpaceae	16	173
			Sciadopityaceae	1	1
			Taxaceae	6	28
		Ginkgoidae	Ginkgoaceae	1	1
		Cycadidae	Cycadaceae	1	91
			Stangeriaceae	2	3
			Zamiaceae	8	195
		Gnetidae	Ephedraceae	1	37
			Gnetaceae	1	30
			Welwitschiaceae	1	1
			Total	82	974
B	Gymnosperms	Pinidae		66	589
		Ginkgoidae		1	1
		Cycadidae		11	289
		Gnetidae		3	68
			Total	81	947
C	Gymnosperms	Cycads		10	310
		Ginkgophytes		1	1
		Gnetophytes		3	100
		Conifers		70	615
			Total	84	1026

From amongst the total species that exist today, conifers form the dominant vegetation particularly in the temperate boreal forests of the Northern Hemisphere and alpine forests at high elevations, because of their adaptation to cold and dry conditions. Many conifer species are also of huge economic importance as a timber source worldwide.

The risk assessment, called the Sampled Red List Index for Plants, conducted by plant scientists at the Royal Botanic Gardens, Kew, UK, found gymnosperms to be at the most at-risk group of plants, with more than 75 per cent of cycad species currently threatened with extinction (Gilbert, 2010). According to a rough estimate in 2010 by IUCN there are 1021 described species of gymnosperms, of which 909 species have been evaluated for various threat criteria. In all, 322 species have been listed as threatened in 2010 (http://e-pao.net/epSubPageExtractor.asp?src=education.Science_and_Technology.Biodiversity_of_gymnosperms).

Since gymnosperms are among the most threatened living organisms on the planet, with 40 per cent of their species at high risk of extinction (which is about twice as many as the most recent estimates for all plants *i.e.* 21.4 per cent, Brummitt *et al.,* 2015), their conservation is the main focus of global research. Which species (and areas) should be prioritised in conservation programmes is a recurrent question being worked out. One of the various methods that have been proposed to address the prioritization issue in conservation is the Evolutionary Distinct and Globally Endangered (EDGE) approach (Isaac *et al.,* 2007). The protection of species with the highest EDGE scores would ensure the preservation of key lineages representing unique evolutionary features within gymnosperms. Even the EDGE values are also not complete as they take each species independently and thus, ignore the actual risk of extinction associated with internal branches. Whatever attempts are made, it is important that a group of plants which coexisted with dinosaurs and is now threatened, must be conserved.

General Features

In the long evolutionary history of land plants, few developments like formation of pollen grains as the male gametophyte and seeds as plant propagules, need a special mention. These two dispersules led to the success of both groups of seed bearing plants - gymno- and angiosperms which presently dominate nearly all terrestrial ecosystems. Evolution from flagellated, motile antherozoids to non-motile sperms was a big achievement as it freed plants from the films of water. This dependence on water by other embryophytes/tracheophytes such as Bryophytes and Pteridophytes was observed for antherozoids to swim in order to reach the archegonia which housed the egg. These two groups of archegoniates for this simple reason have been known as 'amphibians of plant kingdom', as they inhabit land but depend on water to complete their life cycle. They also require water for dehiscence of antheridia to release the antherozoids, for archegonia to release mucilage to trap the gametes and for sporogonium dehiscence to release the spores. It has been one of the many reasons for the small size of bryophytes and limitations of pteridophytes to form wood. Formation of seeds also benefited plants as they could disperse propagules to greater distances, increasing chances of survival due to decreased kin competition.

During evolution, gymnosperms became the first successful land plants which produced seeds rather than spores and were no longer dependent on water for fertilization. Another feat was they could broadcast their male gametes through pollination.

Further, both gymnosperm dispersules - seeds and pollen grains, complete development in a protected enclosure, pollen in the pollen chamber and seed in the ovule (covered by perianth or scales). Here they are provided substantial nutrition for proper growth. Thus, plants invest a large resource in their sexual reproduction. The seeds are borne beneath cone scales or other specialized structures and are protected against stress from harsh environmental conditions.

The cone morphologies are consistently associated with specific ranges in size of the seed, the final outcome of reproduction (Andrew *et al.*, 2017). Though in some cases as in *Pinus,* seeds are dispersed by wind, in many others, seeds give cues to the dispersers. In *Gnetum* for example, seeds are brightly coloured and fleshy, thus encouraging animals to disperse them **(Plate III.6).**

The other dispersule – pollen grain or male gametophyte being immobile, needs to be assisted by some agency such as wind or/and animals in its journey to the ovule. It has to be delivered directly to the egg. Pollen is, therefore, enticed after landing on the ovule and is taken down by pollination drop, a unique phenomenon in gymnosperms. The drop is sticky, chemically similar to the nectar of flowers containing about 8 per cent sugar (glucose and fructose) and traces of amino acids. It is resorbed inside the female cone, carrying pollen grains with it. It is brought to pollen chamber where pollen grains complete their development before embarking on fertilization. The spermatogenous cell of the male gametophyte forms two non-motile sperms which are passively carried to the egg by the pollen tube. This siphonogamic type of sperm transfer is typical for all extant seed plants, with the exception of cycads and *Ginkgo*, which have multiflagellated, swimming spermatozoids that are released from the bursting pollen grain in vicinity of the archegonium. Although, in the cycads and *Ginkgo* pollen tube is formed but it is haustorial in nature, deriving nutrition from the surrounding tissues as it grows (Choi and Friedman, 1991). Siphonogamy is also associated with formation of nucellar beak and pollen chamber in gymnosperms. The former is probably of not much phylogenetic significance while the latter plays an important role in reproduction of this group of seed plants.

Heterospory, formation of different spores by male and female plant is more evolved than homospory. It is quite distinct in gymnosperms, and occurrence of unisexual 'cones' is perhaps the first visible expression of difference in male and female reproductive organs. The megaspore is borne in the nucellus of the ovule while microspores arise in the microsporangium; the number of microspores being much more than the female counterpart.

Another feature which has helped gymnosperms to dominate several ecosystems is the occurrence of secondary growth that has equipped these plants with appropriate supporting and conducting tissues. This has enabled naked seeded plants to produce exceptionally large and strong trees *e.g.*, sequoias. Such differentiation of vascular tissues also helped plants get well-structured and complex developmental patterns besides having contributed to their longevity and robustness. In conifers it has also contributed to various by-products like resins and turpentine with significant commercial value.

Gymnosperms evolved in the Devonian period (of the Paleozoic era) from progymnosperms which exhibited formation of wood and heterospory, but which reproduced by releasing spores. The Mesozoic, which followed Paleozoic was known as ERA OF DOMINANCE OF GYMNOSPERMS. The non- flowering seed plants persist today as a diverse group of about 1050 species of trees, shrubs, or climbers. They are found growing throughout the world, in most ecosystems, and are particularly prominent in cool temperate forests.

Common examples of living gymnosperms include cycads (*Cycas, Dioon, Zamia*), maidenhair tree (*Ginkgo*), cedar (*Cedrus*), fir (*Abies*), juniper (*Juniperus*), pine (*Pinus*), redwood (*Sequoia, Metasequoia, Sequoiadendron*), spruce (*Picea*), and Mormon tea or joint fir (*Ephedra*). Most of these plants are trees with great economic and ecological importance. Conifers which include in general pines, firs, spruces, hemlocks, redwoods, cedars, cypress, yews, and several Southern Hemisphere genera are the best known and are trees with great economic value attached because of the wood produced.

Many studies have been conducted on the phylogenetic inter- and intra-relationships of gymnosperms and frequent comparisons are made with flowering plants. Large-scale molecular analyses have indicated, that any one of the living gymnosperm groups did not give rise to flowering plants. Instead, living gymnosperms form a clade that is a sister group to flowering plants. Though there exists uncertainty about relationships amongst members of the clade gymnosperms, most analyses place cycads at the base of the gymnosperm tree because they possess a series of ancestral features that are fern-like. *Ginkgo* with a number of ancestral traits, and its similarities with conifers has its own position, has been placed separately. Conifers make up the bulk of gymnosperm species while gnetophytes, a distinctive puzzling group has created a great deal of debate among botanists. They possess certain vegetative and reproductive traits amazingly similar to flowering plants, but recent molecular analyses have classified them in the gymnosperm lineage.

Several distinct features of gymnosperms are summarized:

- Gymnosperms are naked seeded plants with novel features in their life cycle.
- They are an ancient group of seed plants with a few shared characters probably due to its divergence times. Although phylogenies point toward monophyly, some features are common to all lineages (cycads, *Ginkgo*, conifers and Gnetophytes), such as outcrossing breeding systems, long generation and reproductive cycle times, and presence of three cell generations in seed: (a) Parent sporophyte in the form of seed coat/coats and persistent nucellus, (b) Female gametophyte which stores food is the present generation, and (c) Future sporophyte in the form of embryo.
- Most plants are evergreen and woody. They are xerophytic, with sunken stomata and thick cuticle **(Plates I.4, V.4)**.
- Living gymnosperms are distributed worldwide (excepting Antarctica), with a majority, particularly the conifers, in temperate and sub-arctic regions.
- Gymnosperms include many of the largest trees *e.g., Ginkgo* can grow very large. Among the conifers, *Sequoiadendron giganteum* is the largest of all trees, but many other giants exist as well.

- Wood and leaf, habit and habitat, physiology and anatomy - all are quite variable between the four major groups.
- Gymnosperms are anatomically and morphologically complex. The anatomy of most gymnosperm wood is extremely regular which is partly because of absence of vessels. In many taxa, including a majority of conifers and *Ginkgo*, tracheids produced in the spring are larger in diameter, while those of the autumn wood are smaller in diameter and have thicker walls. The seasonal variation in tracheid size results in annual rings in the wood.
- Many trees have annual rings, permitting approximate determination of the age of an individual, but in harsh environments many years may pass before the first ring is made. For example, Lertzman (1990) found mountain hemlock seedlings up to 18 years old (based on bud scar counts) that had yet to form their first ring. The greatest age known from ring counts is about 5000 years, for *Pinus longaeva*. Several other species, such as *Thuja plicata* and *Taxus baccata*, might achieve comparable ages but as they live in moist climates where the tree's woody heart rots away while the rest of the tree is in the prime of life, there is no record of the rings left behind.
- Many other trees, including most tropical species, and all non-tree gymnosperms (Cycadales and Gnetales), do not have regular annual rings. Age estimates for such plants are usually based on extrapolation of observed growth rates. Such extrapolation may lead to estimation errors of several hundred percent (usually overestimation, because trees grow more slowly as they age).
- In addition to presence of seeds and pollen, gymnosperms are characterized by a number of vegetative features such as presence of dimorphic branches – long and dwarf shoots and dimorphic leaves – scale and foliage.
- These plants have long life spans and long reproductive cycles characterized by a delayed gametogenesis that usually ends until pollination. For example, gametes are not present in pollen when it is shed, they are formed inside the ovule. Archegonial differentiation and egg maturation is delayed till pollen contacts nucellus.
- Most of the gymnosperms are dioecious and in cases where they are monoecious, the male and female branches are different as in *Pinus*. Bisporangiated cones have been reported for gnetophytes and some conifer species. In gnetophytes, abortive ovules are often found in the collars of male cones therefore, these plants are referred to as gynodioecious **(Plate III.5)**. However, because ovules in such cases are sterile, they could also be referred to as dioecious plants (Endress, 1966).
- Reproductive structures are called cones due to their conic shape as in conifers or strobilla as in others.

- The female is the ovulated (seed) cone or the megasporangiated strobilus, while the male is the pollen cone or microsporangiated strobilus. More often conifer reproductive units are referred to as cones while in others they may be referred to as strobilus. Cones (or strobili) are considered simple when the structures (microsporangium or megasporangium) that bear gametes are directly attached to the central axis of the cone as in *Pinus*, or compound when the gamete bearing structures are borne in the axil of a bract, making up a clustered complex as in *Gnetum* **(Plate III.3).**
- Female structures could be simply a pair of naked ovules as in *Ginkgo* or complex structures as scale-seed-complex in conifers. Morphological differences have been interpreted as adaptations to seed dispersal, seed protection and also pollination mechanisms.
- In conifers with male and female structures borne by the same plant, endogamy is avoided through different mechanisms such spatial or phenological separation of female and male cones. In many conifer species male cones are restricted to lower branches whereas the female cones are formed only at highest branches (also helping seed dispersal by air). Phenological separation includes species where young tree produces only pollen cones and older trees ovulated cones and species in which distinct reproductive structures mature at different seasons.
- Anemophilous pollination is common, but pollen transportation by insects is present in atleast two lineages. Pollination mechanisms involve secretion of pollination drop and distinct pollen morphologies.
- In *Ginkgo* and cycads, sperm cells are motile, a feature unusual for seed plants. The two lineages in gymnosperms show zooidogamy, characteristic of spore-bearing plants such as ferns. Motility in gametes of gymnosperms is not fully useful, since spermatozoids are enclosed and transported by pollen grains and thereafter by pollen tube. Probably retention of this character is correlated with haustorial development of microgametophytes in these groups.
- Besides modifications in ovule bearing structures, pollination mechanisms are diverse in gymnosperms. Characters associated with pollination, such as ovule direction, production and retraction of a pollination drop **(Plates II.6, V.13)** and pollen shape are variable even between species from the same genus. Pollen grains in gymnosperms are larger than in angiosperms, though they are light due to low water content. They can therefore, be dispersed by air through long distances, and in some cases assisted by wings or aerial sacs, which are expansions of outer wall (exine). Wind pollination requires elevated pollen production to ensure that atleast some pollen grains reach female cones.
- Female gametogenesis consists of a coenocytical megagametophyte development and its cellularization thereafter occurs by alveoli formation. Endosperm is a pre-fertilization tissue in gymnosperms.

- Female gametogenesis is similar between groups, whereas male gametogenesis is variable as expressed in number of prothallial cells in a pollen grain. Male gametogenesis involves meiosis to form tetrads of microspores and distinct number of subsequent mitosis before sperm differentiation. Sperm could be, with or without mobility.
- Differentiation of two or more archegonia is a post pollination development, exception being *Gnetum* where archegonia are absent. Wherever present, archegonia lack neck canal cells.
- There is evidence of genetic post fertilization systems to avoid the development of autogamous embryos as in *Pinus*. Here, a genetic overdominant lethal system stops the development of embryos produced by self fertilization (Williams *et al.*, 2003).
- In absence of pollination drop different mechanisms allow to capture pollen and stimulate germination. In some cases, the ambient humidity accumulated between micropyle and cone central axis works as pollen drop, capturing pollen and allowing germination. In *Abies, Welwitschia* and some species of Araucariaceae, growing of nucellus beyond micropyle to reach pollen has been observed. In other cases, a delayed secretion has been observed [*Pseudotsuga* and *Larix* (Pinaceae)], where the secretion is more related with fertilization mechanisms rather than with pollination.
- A time gap exists between pollination and fertilization, ranging from one to several weeks to several months as in cycads.
- Double fertilization has been reported in *Gnetum* sp. and *Ephedra*. It occurs in two different ways, since archegonia could be present (*Ephedra*) or absent (*Gnetum* and *Welwitschia*).
- In gymnosperms polyembryony is feasible to occur by two different processes: "simple polyembryony" and "cleavage" or "monozygotic polyembryony".
- In all gymnosperms, polyembryony is common and results in a selection of a single embryo from a larger number, long before the seed is matured. The suspensor of the gymnosperm embryo is an organ of competition, which in *Ginkgo* is short. It seems to be the only gymnosperm which occasionally matures more than one of its several embryos. It is much more infrequent in other gymnosperms. Simple polyembryony is found among cycads, with several archegonia. Among conifers there is peculiar splitting of the embryo especially rosette embryos cleavage, while in *Ephedra* cleavage is of regular occurrence (Buchholz, 1920).
- Embryo bears two as in *Cycas* or more (10-11 in *Pinus*) cotyledons.

★*Functionally unisexual reproductive structures, unitegmic ovules, presence of three generations in seed, and anemophilous pollination syndromes are common features for the group.*

Chronology of Gymnosperm Evolution

Although molecular studies have done much to improve our understanding of evolutionary relationships between extant gymnosperms, changing relationships at evolutionary time-scales are best elucidated through the fossil record **(See Box 1.1;** adapted from Christopher J. Earle 2018 – The Gymnosperm Database**)**.

Box 1.1

- Late Devonian (385-359 million years ago [my]): Earliest seed plants arise (Hill, 2005).
- Middle Pennsylvanian (310 my): Earliest conifers, cycads, and *Ginkgo* arose at about the same time, perhaps from a precursor in the Cordaitales, a plant that bore seeds in cone-like structures. These earliest gymnosperms were not referrable to any families now existing. The Cordaitales persisted until the mass extinction at the end of the Triassic (201 my) (Hilton, 2014).
- Permian (299-251 my): First appearance of Ginkgoales. The group is most diverse and abundant during the Jurassic (Thomas and Spicer, 1986 in Hill, 2005).
- Lower Permian (299-271 my): More recognizable cycads appear, *e.g., Crossozamia* (Gao and Thomas, 1989); cycad diversity and abundance peaked during the Mesozoic (Hill, 2005).
- Middle Permian (271-260 my): First evidence of Gnetophytes, limited to pollen evidence (Crane, 1988); macrofossils attributable to Gnetophytes appear only much later (Cretaceous).
- Upper Triassic (235-202 my): First appearance of the Bennettitales. This was one of the most abundant and diverse gymnosperm groups in the Mesozoic, although it became extinct (along with the dinosaurs) at the end of the Cretaceous. Evolutionary relationships are unclear but the Bennettitales may have been most closely related to the Gnetales (Hill, 2005).
- Triassic, earliest Norian (228-225 my): The first appearance of Podocarpaceae, the extinct species *Rissikia media* (Townrow, 1967 in Rothwell *et al.,* 2012).
- Lower Jurassic (202-176 my): First appearance of *Ginkgo*, the oldest extant conifer genus (Tralau, 1968).
- Jurassic, Sinemurian (197-190 my): First record of Cupressaceae, the extinct species *Austrohamia minuta* (Escapa *et al.,* 2008 in Rothwell *et al.,* 2012).
- Jurassic, Pliensbachian (190-183 my): First record of Taxaceae, the extinct species *Paleotaxus rediviva* (Nathorst, 1908 in Rothwell *et al.,* 2012).

- Jurassic, Pliensbachian (190-183 my): First record of Araucariaceae, the extant genus *Araucaria* (Arrondo and Petriella, 1980 in Rothwell *et al.,* 2012).
- Jurassic, Kimmeridgian (156-151 my): First record of Pinaceae, the extinct species *Eathiestrobus mackenziei* (Rothwell *et al.,* 2012).
- Lower Cretaceous (130 my): First appearance of *Pinus*, the extinct species *Pinus yorkshirensis* from a fossil cone preserved in the Wealden Formation in Yorkshire, UK (Ryberg *et al.,* 2012).
- Cretaceous, Aptian (125-112 my): First Gnetales macrofossils, in the form of *Drewria potomacensis*, an extinct species with leaves resembling those of *Gnetum* (Crane and Upchurch, 1987 in Hill, 2005).
- Cretaceous, Conacian (89-86 my): First record of Sciadopityaceae, the extant genus *Sciadopitys* (Ohsawa *et al.,* 1991 in Rothwell *et al.,* 2012).

Do You Know?

*The largest of all, a giant sequoia (**Sequoiadendron giganteum**), is 55,040 cubic feet, followed closely by the largest coast redwood (**Sequoia sempervirens**) at 36,890 cubic feet.*

*The largest sugar pine (**Pinus lambertiana**) is 8,990 cubic feet. The reason why sugar pines are the biggest pines is because they have very little taper. They shoot up out of the ground like almost perfect cylinders, sometimes with all of their crown in the top 20 per cent or so of the tree. For all tree species, the key to achieving high wood volume is a big diameter, sustained to great height.*

***Sequoia sempervirens**, which is only one of many gymnosperms that can sprout from fallen stems or cut stumps. Some cycads produce clones by producing buds at the base, which may become separated from the main plant and then root. The oldest gymnosperm clones are likely to be found among species prone to vegetative reproduction in an area of extraordinarily stable climate, such as cycads of the genus Encephalartos growing in South Africa.*

***Parasitaxus usta** belonging to family Podocarpaceae is the only parasitic gymnosperm known so far. It is purple due to high anthocyanins in vacuolar sap. It lacks roots and parasitizes Falcatifolium taxoides.*

Are Gymnosperms Hydrophobes?

Several classical textbooks, have referred to gymnosperms as the only seed plants without aquatic species. According to global survey of Kozlowski *et al.*, (2015) about 18 per cent of all gymnosperm species are in fact wetland inhabitants and that almost 1 per cent of them are perfectly adapted to aquatic habitats and are therefore, hydrophilic. Eighty two percent of the extant species (805) of gymnosperms are clearly hydrophobic (leaving only a handful species of hydrophilic existence). The reason for this hydrophobicity needs attention since gymnosperms existed on earth for more than 300 million years and as an extremely successful plant group in the past (Ran *et al.*, 2010). The existence of at least one of the potential ancestors of gymnosperms, the fossil genus *Cordaites*, which presumably was a freshwater paralic swamp taxon with stilt roots, has made the situation more complex (Falcon-Lang and Bashforth, 2005; Raymond *et al.*, 2010). Scientists speculate that more fossil aquatic gymnosperm species must have existed in past (Sculthorpe, 1967). This is because modern gymnosperms represent only a small taxonomic and ecological fraction of what once dominated Mesozoic flora (Williams, 2009).

Many of the hydrophilic gymnosperms described by the workers are at the brink of extinction. Notably, the only obligate aquatic gymnosperm, *Retrophyllum minus*, possesses only four localized natural populations on New Caledonia, and the poorly studied *Glyptostrobus pensilis* is probably extinct in wild in China, and extremely rare in Laos and Vietnam (Debreczy and Racz, 2011; Thomas *et al.*, 2011).

The reasons for hydrophobicity in extant gymnosperms has been discussed by several groups. The most visible and widespread phenomenon of buttressing (stem hypertrophy) shown by all hydrophilic species provides an explanation to hydrophobic nature. The process of stem hypertrophy is stimulated by ethylene, causing an increase in porosity of the tree base and thereby enhancing aeration (Cronk and Fennessy, 2001). The two extant species *Taxodium distichum* and *G. pensilis* show in addition, 'root knees', which in turn improve their adaptive capacity to wet environments **(Fig. 1.1)**. In addition, they have also developed specific reproduction strategies such as seed dispersal by water and germination in water for their life in aquatic conditions (Earle, 2013). The hydrophobia of a large majority of gymnosperms might therefore, be reflective of their more rigid life form and ancient reproduction strategies (Williams, 2009).

* ***Glyptostrobus pensilis***, also known as **Chinese swamp cypress**, is the sole living species in the genus *Glyptostrobus*. It is native to subtropical south eastern China, from Fujian west to south east Yunnan, and also grows in northern Vietnam. It typically grows in river banks, ponds and swamps, growing in water up to 60 cm (24 in) deep. Like the related genus *Taxodium*, it produces 'cypress knees', or pneumatophores, when growing in water are believed to help transport oxygen to the roots.

The ability of cypress knees to allow easy diffusion of O_2 from the atmosphere to the attached submerged root is supported by a number of past ecomorphological observations.

Figure 1.1: *Taxodium distichum* with Cypress knees.

A majority of gymnosperms, and in particular conifers, suffer from low flooding tolerance with decreased root and needle growth, root decay, and with the formation of abnormal wood and bark tissues (Yamamoto *et al.,* 1987; Kozlowski, 1997). The role of tracheids can not also be ignored. Gymnosperm tracheids are generally more resistant to solute flow than angiosperm vessels contributing to a less efficient transport system. For several such reasons therefore, maximum growth rates of gymnosperms are lower on open, sunny sites and well-watered habitats as compared with those of angiosperms (Bond, 1989). The very narrow tracheids and weakly vascularized evergreen leaves (which are most obvious in conifers) of gymnosperms make them much more adapted to water deficiency and conservation, and coevally less suited to invade waterlogged soils (Bond, 1989; Sperry *et al.,* 2006).

Thus, gymnosperms form a specialized group of plants which due to rigid reproduction strategies, morphological plasticity (Burger, 1981; Stebbins, 1981) has become perfectly adapted to terrestrial life. They are totally unfit to very dynamic aquatic habitats with abrupt changes in water level or rapidly flowing water. The absence of pollination vectors, widespread in angiosperms, might be thus, another argument for the inferiority and absence of gymnosperms in wet environments (Williams, 2009).

Gymnosperms *vs* Angiosperms

Though gymnosperms and angiosperms are considered sister groups by several workers, yet the significant differences between the two has kept the role of gymnosperms in origin of flowering plants a mystery **(Table 1.2)**.

Table 1.2: Gross characters of gymnosperms are compared with those in angiosperms.

Character	*Gymnosperm (EXTANT)*	*Angiosperm*
Habitat	Restricted, very few near water.	Wide range, many are aquatic, submerged.
Habit	Mostly trees, shrubs or rarely vines. One or two parasitic.	Wide range — trees, shrubs, herbs, climbers, parasites, epiphytes.
Morphology	Variation and Specialization not seen.	Stem, root and leaf modifications seen.
	Dimorphism of stem and leaf. No flowers.	Great range in shape and size of stem and leaf, and colour of flowers.
Anatomy	Phloem lacks companion cells.	Companion cells present.
	Xylem lacks vessels (except in Gnetophytes).	Vessels well developed in most species.
	Annual rings generally not demarcated as plants are very slow growing, some with secondary growth not distinct.	In dicots where secondary growth occurs rings are demarcated though monocots do not show because of absence of secondary growth.
Reproduction	Cones or strobili are the reproductive structures which are generally unisexual.	Flowers are formed – bisexual, unisexual or they may be neuter as in capitulum.
	Microspores show prothallial cells.	No such structures associated with ovules.
	Megasporangium is less developed.	Is a highly developed structure.
	Pollination mainly by wind rarely by insects.	Diverse vectors – wind, insects, animals, bats, birds .
	Pollination drop formed .	Nectar and various other devices to capture pollen.
	Pollen and archegonial chambers formed.	No such structures associated.
	Archegonia present which bear egg and ventral canal cell.	Eggs are borne directly inside embryo sac.
	Pollen falls directly on micropyle.	Pollen contacts the stigma first.
	Endosperm is haploid and pre fertilization product.	It is triploid and post fertilization tissue.
	Double fertilization only in *Gnetum* sp. and *Ephedra*.	Double fertilization and triple fusion is basic to sexual reproduction.
Embryogeny	Polyembryony through cleavage and simple is frequent.	Polyembryony of various types and apomixis common.
Seed	Is naked with protection from bracts and scales only.	Seeds are protected inside the fruit.
	Dispersal mechanisms not many.	Dispersal varied and complex and is a syndrome.

Out of all the four lineages of gymnosperms, conifers have resisted angiosperm expansion better than other lineages. Several hypothesis have been proposed for the rise of angiosperms and decline of gymnosperms. According to 'reproductive co-evolution' hypothesis, gymnosperms were outcompeted by angiosperms as wind pollination in gymnosperms became inefficient and speciation was slow. In the carbon framework, it is hypothesised that angiosperm species outcompeted the formerly dominating gymnosperms because former had an improved ability to grow fast (Bond, 1989; Coiffard *et al.*, 2012). This increased growth rate perhaps was because of a more efficient vascular hydraulic network. Further, water transpiration is essential to the photosynthetic process (Brodribb and Feild, 2010) and the seed plants of Jurassic Period (201–145 my) had fairly low photosynthetic capacities (Brodribb and Feild, 2010; Feild *et al.*, 2011).

Classification

Early Views

The early classification of gymnosperms which has drawn attention, placed them as a co-ordinate group with angiosperms within Phanerogams or Spermatophytes.Hooker (1857) placed them between Dicotyledons and Monocotyledons. In 1936, Engler made Gymnospermae a subdivision under which were seven groups (orders):

1. Cycadofilicales
2. Cycadales
3. Bennettitales
4. Ginkgoales
5. Corditales
6. Coniferales
7. Gnetales

Sahni in 1920 gave an elaborate classification in which gymnosperms were spread out under various subclasses.

Class Gymnospermae with four subclasses was organized with various orders - selected have been mentioned below:

Subclass 1 - Pteridosperminae with order Cycadofilicales

Subclass 2 - Phyllosperminae with order Cycadales

Subclass 3 - Stachyosperminae with orders Cordaitales, Ginkgoales and Coniferales

Subclass 4 - Chlamydosperminae with orders Ephedrales, Welwitschiales and Gnetales

Pant (1957) constituted three divisions with following classification:

Division 1 - Cycadophyta with four classes of which class 2 - Cycadopsida had order Cycadales

Division 2 - Chlamydospermophyta with one class – Gnetopsida and two orders Gnetales and Welwitschiales

Division 3 - Coniferophyta with four classes out of which, class 1 - Coniferopsida had three orders – Cordaitales, Coniferales and Ginkgoales, while class 2 - Ephedropsida had single order Ephedrales.

It may be summarized that earlier classifications recognized two major groups in Gymnosperms – Cycadophytes and Coniferophytes.

Molecular and Phylogenetic Basis

The gymnosperms has long been recognized as a distinctive group of plants, but the relationships of the principal subclasses within the gymnosperms are still being debated upon and are being worked out with modern tools of bioinformatics. Whether these subclasses share a common ancestor is one side of the mystery, while the other remains that if so, whether the common ancestor is or is not shared with other group of seed bearing flowering plants, the angiosperms. If all the gymnospermic subclasses share a common ancestor, distinct from the flowering plants, the naked seeded plants – gymnosperms are a unique lineage of plants; otherwise, the term "gymnosperm" has no taxonomic significance **(Fig. 1.2).**

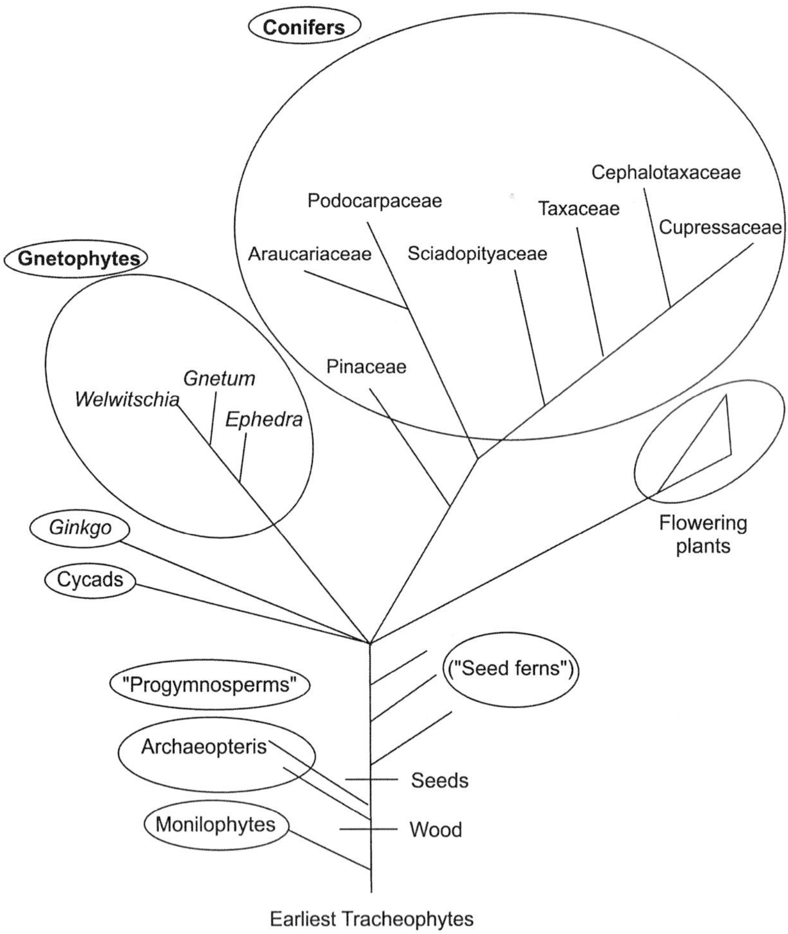

Figure 1.2: A cladogram showing relationships among seed plants. All lineages emerge from a single point represented by earliest tracheophytes (http://www-plb.ucdavis.edu/courses/bis/1c/text/Chapter24nf.pdf).

However, the term "gymnosperm" has a practical significance because it refers to a relatively small and highly distinctive assemblage of plants with ancient lineage, which have been reduced to a largely relictual distribution associated with the evolution of angiosperms (flowering plants). The angiosperm-enriched communities made wind pollination inefficient or limited impacting speciation. However, the cultural and ecological importance of gymnosperms at a global scale can not be ignored.

In a taxonomic debate two major hypothesis have been put forth. In the "anthophyte" hypothesis, the anthophytes are defined to be a clade including both angiosperms and gnetophytes. This calls for a common ancestor for gnetophytes-*Ephedra*, *Gnetum* and *Welwitschia*, and the angiosperms. The other is gnetifer (or the variant 'gnepine,' gnetophytes plus Pinaceae) hypothesis. According to this opinion, conifers (which are usually, but not always, assumed to share a common ancestor with the cycads and *Ginkgo*) and gnetophytes belong to the clade gnetifer. Both the opinions have generated support by various groups of workers and the taxonomic debate for classification of gymnosperms has continued. Though accumulation of molecular data seems to be swinging in favour of the gnetifer or gnepine hypothesis, the recent studies on morphology support the anthophyte hypothesis (Friis *et al.*, 2007).

Christenhusz *et al.* (2011), while reviewing studies available to date, simply assign each of the four gymnosperm groups to a subclass under the embryo bearing plants, Equisetopsida (Class). They also note that under the gnepine hypothesis it would be appropriate to put the gnetophytes and Pinaceae in one subclass, and all other conifers in another subclass.

Earle (2018) who has prepared 'The Gymnosperm Database' believes that though different classification systems exist *e.g.*, Reveal (1998), the classification proposed by Christenhusz *et al.* (2011) is well supported by both molecular and morphological evidence. It does not however, assume validity of the gnepine hypothesis.

Hori *et al.* (1985) were the first to indicate that extant gymnosperms were sister to the angiosperms. However, only three genera, *Cycas, Ginkgo* and *Metasequoia* were included in this study and results were therefore, inconclusive. Subsequent effort was made by Troitsky *et al.* (1991) who used ribosomal RNA and found extant gymnosperms to be monophyletic. The study could not give clear evidence as sampled data was thin with only a total of 11 genera used (three cycads, two gnetophytes, *Ginkgo* and five conifers). Hasebe *et al.* (1992) proved the monophyletic origin of gymnosperms using plastid rbcL sequences on a small set of taxa. The first broadly sampled molecular phylogenetic study to examine gymnosperm relationships was that of Chaw *et al.* (1997). It indicated that extant gymnosperms were sister to the angiosperms. Molecular studies by Ran *et al.* (2010) have demonstrated monophyly for both groups of extant seed plants. Chase and Reveal (2009) recognize each of the four extant groups and also angiosperms as subclasses of class Equisetopsida. This view does not favour separation of cycads, *Ginkgo*, gnetophytes and conifers as a whole. They also differ from others who have proposed individual groups within conifers such as pines, Araucariaceae and Cupressaceae, Taxaceae and Podocarpaceae, individual classes, namely Cycadopsida Brongn., Ginkgoopsida Engl., Gnetopsida Eichler ex Kirpotenko, Pinopsida Burnett (Coniferopsida), AraucariopsidaBobrov and Melikian, Taxopsida Lotsy and Podocarpopsida Doweld and Reveal (cf. Sporne, 1965; Bierhorst, 1971; Doweld and Reveal, 1999; Melikian and Bobrov, 2000; Bobrov and Melikian, 2006). Traditionally cycads, *Ginkgo*, and Araucariaceae with long fossil history have been considered 'primitive' and extant members of these three groups resemble the fossils (Kunzmann, 2007; Sun *et al.*, 2008; Passalia *et al.*, 2010). Gnetidae were often considered as the sister group of angiosperms (Chase *et al.*, 1993), and *Gnetum* L. bears resemblance to an angiosperm in general habit. For this reason, Araucariaceae were often placed at the beginning of gymnosperms, whereas Gnetidae were placed at the end.

Although the extant taxa listed under gymnosperms are clearly monophyletic, their relationships to diverse groups of fossil gymnosperms and other plants remain obscure and incompletely understood. To classify all taxa of gymnosperms, living and extinct is therefore, a difficult task. Based on the work of Chase and Reveal (2009) who adopted 'The Angiosperm Phylogeny Group (APG)' classification, a new system of placing taxa according to phylogeny has been proposed by Christenhusz *et al* (2011). They have recognised the four extant groups as subclasses under class Equisetopsida **(Table 1.3).** In this system, the exact position of Gnetidae with respect to the other subclades of extant gymnosperms has been indicated as problematic and controversial. This is because studies have indicated them to be embedded in the conifer clade as sister to Pinaceae (the so-called gnepine hypothesis: Qiu *et al.*, 1999).

Table 1.3: Classification of Equisetopsida [embryophytes]. Modified after Chase and Reveal (2009).

Class Equisetopsida C.Agardh, Classes Pl.: 7. 20 May 1825
[Bryophytes]
Subclass Anthocerotidae Engl., in H.G.A. Engler and K.A.E. Prantl, Nat. Pflanzenfam. I, 3: 1, 6. 10 Oct 1893.
Subclass Bryidae Engl., Syllabus: 47. Apr 1892.
Subclass Marchantiidae Engl., in H.G.A. Engler and K.A.E. Prantl, Nat. Pflanzenfam. I, 3: 1, 5. 10 Oct 1893.
[Lycopodiophytes]
Subclass Lycopodiidae Beketov, Kurs Bot. 1: 115. 1863.
[Monilophytes] ferns
Subclass Equisetidae Warm., Osnov. Bot.: 221. 22–28 Apr 1883.
Subclass Marattiidae Klinge, Fl. Est-Liv-Churland 1: 93. 22–28 Jun 1882
Subclass Ophioglossidae Klinge, Fl. Est-Liv-Churland 1: 94. 22–28 Jun 1882.
Subclass Polypodiidae Cronquist, Takht. and Zimmerm., Taxon 15: 133. Apr 1966.
Subclass Psilotidae Reveal, Phytologia 79: 70. 29 Apr 1996.
[Gymnosperms]

Subclass Cycadidae Pax 1894
Subclass Ginkgoidae Eng. 1897
Subclass Gnetidae Pax 1894
Subclass Pinidae Cronquist, Takht. & W. Zimm. 1966
[Angiosperms]
Subclass Magnoliidae

According to Christenhusz *et al.* (2011), the extant gymnosperms can be divided into four subclasses under class EQUISETOPSIDA.

The outline (broad) of classification followed in the book is:

SUBCLASS I – CYCADIDAE

ORDER A – CYCADALES

Family 1 – Cycadaceae-1 genus, ca 107 species

Family 2 – Zamiaceae-9 genera, ca 206 species

SUBCLASS II – GINKGOIDAE

ORDER B – GINKGOALES

Family 3 – Ginkgoaceae-1 genus, 1 extant species

SUBCLASS III – GNETIDAE

ORDER C – WELWITSCHIALES

Family 4 – Welwitschiaceae-1 genus, 1 species

ORDER D - GNETALES

Family 5 – Gnetaceae-1 genus, 30 species

ORDER E – EPHEDRALES

Family 6 – Ephedraceae-1 genus, ca 40 species

SUBCLASS IV – PINIDAE

ORDER F – PINALES

Family 7 – Pinaceae-11 genera, ca 225 species

ORDER G - ARAUCARIALES

Family 8 – Araucariaceae-3 genera, ca 41 species

Family 9 – Podocarpaceae-19 genera, ca 180 species

ORDER H – CUPRESSALES

Family 10 – Sciadopityaceae-1 genus with a single species

Family 11 – Cupressaceae-29 genera, ca 130 species

Family 12 – Taxaceae-6 genera, ca 28 species

Chapter 2

Subclass Cycadidae

Based on the fossil record, cycads are believed to have appeared in the carboniferous period, approximately 300 million years ago. Later they dominated the Mesozoic forests along with Coniferales and Ginkgoales.

Cycads (early non-flowering seed plants) exhibit a number of characteristics that reflect their evolutionary position between ferns (non-seed plants) and angiosperms (flowering seed plants). Such characteristics include pollen tubes, which release motile spermatozoids before fertilization (as in ferns); dichotomous branching (versus axillary branching in higher plants); and ovules, that are borne on the margins of leaf-like megasporophylls. The large, free-nuclear megagametophytic stage is a pre-fertilization structure. These characteristics among others place cycads at a key node in plant evolution.

The most primitive extant gymnosperms, the cycads contain (by tally of Whitelock, 2002) 289 species in three families:

1. Cycadaceae (91 spp.)
2. Stangeriaceae (3 spp.)
3. Zamiaceae (195 spp.) (The gymnosperm database, www.conifers.org/zz/Cycadales.php). However, Christenhuz *et al.* (2011) recognize only two families—Cycadaceae and Zamiaceae.

Due to their largely tropical distribution and tendency toward endemism, some species are undescribed, while others, doubtless, or undiscovered. The systematics of some genera are poorly understood (Jones, 1993; See also Stevenson, 1999) and scientists need to carry out extensive work on this group of plants.

The plants are woody, long-lived, and unisexual. The main root is thickened, fleshy, and often tuberous. All species have coralloid roots that are N_2-fixing due to symbiotic blue-green algae; many species also form vesicular-arbuscular mycorrhizae (Jones, 1993; Brundrett, 2008). Stems may be underground or emergent. In most species, leaves are once-pinnate and form a palm-like crown. Leaves of *Stangeria* are fern-like, while those of *Bowenia* are bipinnate. Reproductive structure is a large strobilus. The Cycadales resemble Ginkgoales as both orders are unique among seed plants in having motile spermatozoids (Jones, 1993).

Cycad stems are succulent with a structure similar to monocots and gradually enlarge as the plant ages. Branching, when it occurs, is generally adventitious, and bulbils or suckers may be required to vegetatively reproduce cycads, either naturally (as in many species of *Encephalartos*) or horticulturally. Cycads have the largest, (up to 3 mm across) apical meristem found in vascular plants.

The plants are distributed in tropical and subtropical latitudes in the Americas, Caribbean, sub-Saharan Africa, Madagascar, eastern India, China, Japan, south east Asia, Oceania and Australia. Some species are frost-tolerant (Jones, 1993). Discontinuous geographical distribution and small area occupancy.

With trunks up to 20 m tall and 50 cm in diameter, *Lepidozamia hopei* is probably the tallest individual cycad. Both *Encephalartos laurentianus* and *Macrozamia dyeri* achieve diameters of 100 cm; the former with heights (or lengths, as it is often decumbent) to 15 m, is the most massive of cycads. Moreover, it often grows in clumps. The greatest diameter 450 mm, though, is attained by *Cycas thouarsii* (Norstog and Nichols, 1997).

Cycadales is the sole order in subclass **Cycadidae** Pax ex Prantl 1894 (http://www.conifers.org/zz/Cycadales.php).

Are Cycads Living Fossils?

- The cycads are gymnosperms that superficially resemble palms or large ferns, but are not closely related to them. Of the four subclasses with their orders that comprise gymnosperms (discussed in following chapters as part of the syllabus), the Cycadales are considered to be the most primitive **(Fig. 2.1).**
- Though modern cycads have the appearance of being ancient, yet they are considered recent arrivals (blogs.discovermagazine.com/./living-fossil-cycad-plants-are-actually-evolutions-com). They have retained the basic shape and form of their long-extinct ancestors, but were not reported to exist there around at the same time.
- Cycads appeared around 280-300 million years ago and dominated in Jurassic and Cretaceous period. However, they got outcompeted by flowering plants and suffered from decline of their dinosaur pollinators. Following this they started to disappear. They enjoyed a second bout of expansion around 12 million years ago.

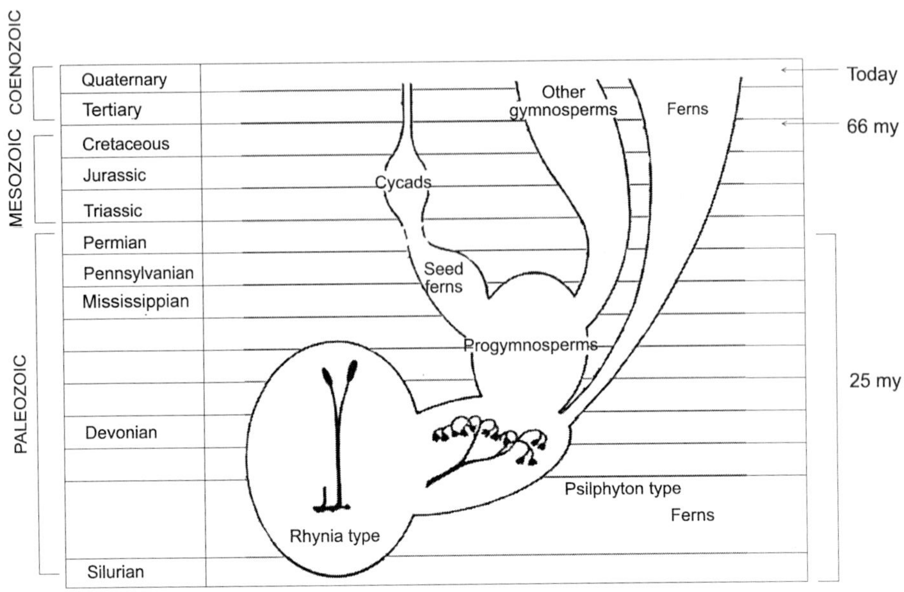

Figure 2.1: Evolution of cycads, gymnosperms and ferns.

- Whatever the cause, the cycad comeback was short-lived and no new species has emerged in the last two million years. The new study using molecular evidences has shown that surviving cycads species are actually not relics of the dinosaur era but the result of an evolutionary explosion among cycads that began about 12 million years ago. About two thirds of the group are now reported endangered.
- According to Nagalingum *et al.* (2011) it would be wrong to call them as 'living fossils' because the living species aren't much older than 12 million years. They may be referred to as only ancient. Most cycads are now threatened, endangered or extinct in the wild. All cycad genera are on Appendix 2 of CITES (the Convention on International Trade in Endangered Species. For the most update species list, the World list of Cycads can be found at: http://www.cycads.org/pages/worldlist.htm.).

Order Cycadales

Family Cycadaceae

Cycadales is an order with families Cycadaceae and Zamiaceae. *Cycas, Ceratozamia, Dioon* and *Zamia* are some common genera distributed in families – Cycadaceae and Zamiaceae and *Cycas* is the only genus included in family Cycadaceae. Cycads are occasionally considered 'living fossils' or evolutionary relicts and are of great scientific and conservation value because of their long evolutionary history.

Cycads have survived through dramatic tectonic activities, climate fluctuations and environmental variations making them of great significance in studying evolution and conservation biology.

Cycas

Distribution

Cycas is a slow growing, long-lived, evergreen and palm-like tree with distinct leafy sporophylls. In Asian region it is distributed in China, India and Japan. Amongst various species *Cycas revoluta*, an ornamental plant is native of China and Japan. It is commonly known as Sago palm. In India, *C. circinalis* is found in India in Malabar hills, Tamil Nadu and Odisha. *C. pectinata* is distributed in Bihar, Assam, and Himalayas. *C. rumphii* is found in Andaman and Nicobar Islands. *C. siamensis* is found in Mayanmar, Sri Lanka and China, while *C. boddomei* is found in eastern Andhra Pradesh and Tamil Nadu.

Sporophyte

Main plant is a sporophyte with root, stem and leaves. Its appearance is palm-like and is an evergreen plant **(Fig. 2.2A).**

Morphology

- Root is long primary tap root which is short-lived and gives rise to lateral roots. The upper part of these becomes modified into apogeotropic coralloid roots arising from the lateral branches of the tap root.
- The stem is tuberous when young but later it becomes thick, columnar, unbranched with an armour of thick persistent leaf bases at maturity. The apex remains surrounded by a crown of large compound leaves.
- Leaves are spirally arranged, and dimorphic. The scale leaves are dry, brown, membranous and protective.
- The foliage leaves are green, and photosynthetic. These leaves develop under the scale leaves and are pinnately compound. The pinnae or leaflets are borne on the rachis.
- The petiole is long and has a spiny sheathing membrane. The pinnae are lanceolate, unicostate with entire margin and the mid vein or the costa ends in a spine. When young, the pinnae remain circinately coiled along the rachis **(Plate I.1).**
- Coralloid roots are produced in clusters **(Fig. 2.2B)** at the base of the stem and protrude out over the ground. They are dichotomously branched and greenish brown in colour. They contain an algal zone in the cortex. This algal zone is inhabited by blue-green algae like *Nostoc* and *Anabaena*, which grow in symbiotic association with coralloid root. They are called coralloid roots due to their coral-like appearance **(Plate I.2).**

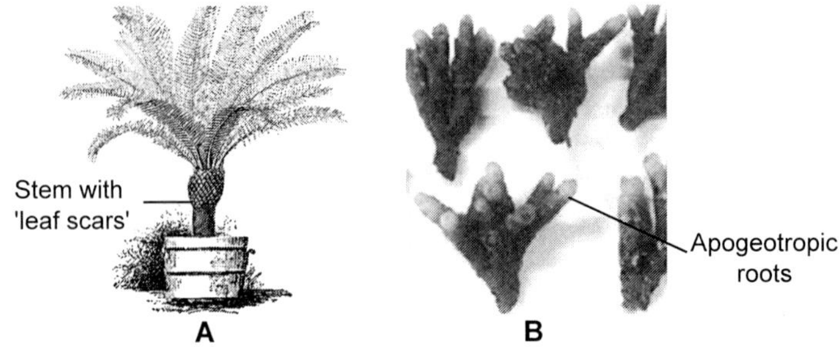

Figure 2.2 A&B: *Cycas* Sporophyte.

A - The sporophyte with an unbranched stem that bears a crown of leaves. B - Specimens of dichotomously branched coralloid roots.

- The symbiotic relationship between cycad root and blue-green algae is common. However, algae do not enter roots in the very beginning. Hence cycad bulbils don't germinate with cyanobacteria in tow but acquire it from the surroundings. To acquire the symbiont, roots begin to form specialized structures called precoralloid roots. Unlike other roots that generally grow downwards, these structures grow upwards and situate themselves in the upper layer of soil where enough light penetrates for cyanobacteria to carry out photosynthesis.
- The cyanobacteria enter precoralloid roots through tiny cracks or cuts and cause a change in root development. *Cycas* then initiates their development into true coralloid roots, with an algal zone. *Cycas* appears to be in full control of the relationship, doling out carbohydrates in return for nitrogen depending on the demands of the immediate environment. Coralloid roots can be shed and reformed throughout the lifetime of the plant.

Anatomy

Stem

- Epidermis formed of thin-walled cells, is incomplete and appears ruptured due to presence of leaf bases. It remains covered with cuticle.
- Cortex is large, parenchymatous, and thin-walled with mucilaginous canals distributed all over and storage starch in cells. There are two girdling traces and two direct traces **(Fig. 2.3)**.
- Endodermis and pericycle are inconspicuous.
- Stele present midway between periphery and centre is ectophloeic siphonostele and comprises small, several closely arranged vascular bundles that are conjoint, collateral, open and endarch. The vascular

bundles are separated by medullary rays. Tracheids in protoxylem have spiral thickenings while those of metaxylem have scalariform thickenings; the secondary xylem elements show bordered pits. Phloem sieve tubes are associated with albuminous cells but lack companion cells.

- Pith is large, parenchymatous with mucilage canals.

STEM is marked by primitive characters such as soft spongy wood which is **manoxylic where stele is thin while cortex and pith are massive **(Figs. 2.3, 2.4A)**. Also seen are leaf traces in the form of girdle and direct ones. The epidermis is interrupted by leaf bases. The mature stem shows several small rings of secondary tissues and therefore, is said to be polyxylic.*

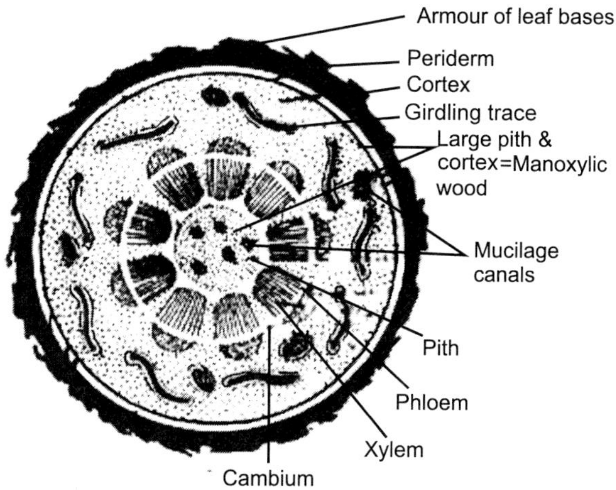

Figure 2.3: *Cycas* T.S. Young Stem.

Diagrammatic sketch showing features of a manoxylic wood. The cortex and pith are massive while the vascular region is narrow.

Secondary Growth

The cambium is short lived and during secondary growth, a ring of secondary tissues is formed with secondary xylem towards inside and secondary phloem outside. The activity soon ceases and after a while another ring of cambium appears and the secondary tissues are formed **(Fig. 2.4B)**. Like this a few short-lived cambial rings arise and patches of secondary tissues are formed till this activity slows down and only patches of **mesarch** xylem are seen towards the end of the activity. In the cortical region also, secondary tissues are formed due to the activity of cork cambium. The secondary tissues are also marked by the medullary rays which appear thin and are filled with sago starch.

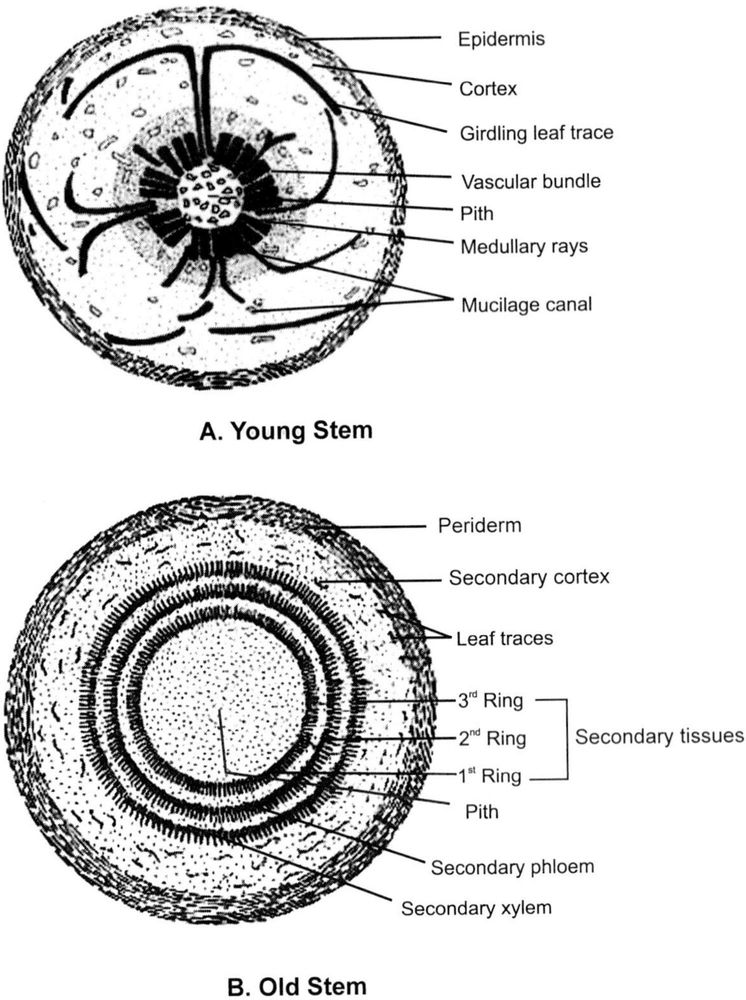

Figure 2.4 A&B: *Cycas* T.S. Stem.

A - Shows girdling leaf traces, a primitive character. Also seen is the cortical region rich in mucilage canals. B - Anomalous secondary growth to show several small rings of secondary tissues.

Root

- The internal structure of the root is somewhat identical with that in angiosperms. Outermost is the single layered epiblema, followed by a large, thin-walled cortex.
- The cortical region contains a number of mucilage canals with a layer of epithelial or secretory cells, as seen in the stem.
- Single layered endodermis with casparian strips and many-layered parenchymatous pericycle encircle the stele, which is diarch (in young) to tetrarch (polyarch in old) **(Fig. 2.5A&B).**

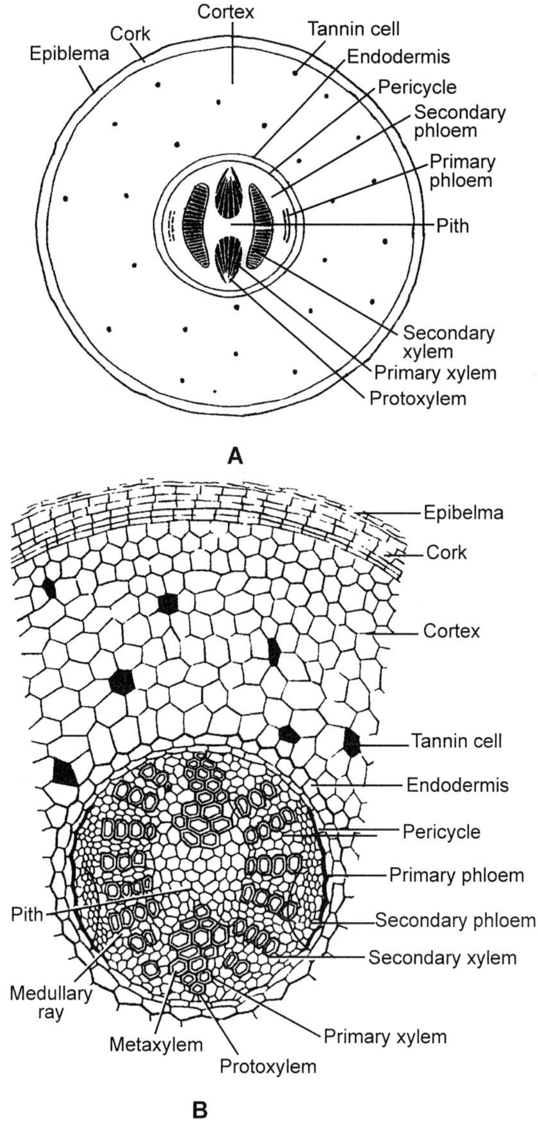

Figure 2.5 A&B: *Cycas* T.S. Root.

A - Outline diagram to show basic structure. B - Cellular details.

- The protoxylem consists of tracheids with spiral thickenings and metaxylem with scalariform and pitted thickenings.
- Phloem in young root consists of sieve tubes and phloem parenchyma but in older roots it consists of bast fibres.
- Secondary growth begins early and follows the same pattern as that of dicot root but is more or less irregular.

- The roots in some cases show two periderm layers, which arise from two layers of phellogen, one in outermost layer of cortex below the epidermis and the second in the pericycle.
- The outer periderm is short-lived as it is peeled off soon after the formation of cork by inner phellogen.

Coralloid Root

- Sometimes during bulbil germination, the cortex gets inhabited by blue-green algae.
- The rapid multiplication of algae forms a conspicuous blue-green algal zone midway between the vascular bundles and the epiblema.
- The algal zone may sometimes even contain diatoms, *Oscillatoria, Azotobacter, Pseudomonas* besides *Nostoc* and *Anabaena* commonly encountered in algal zone **(Fig. 2.6A&B; Plate I.3).**
- These roots curve upwards, branch dichotomously and profusely, forming coralloid masses above ground. They are known as corallorrhiza.
- The vascular structure is about the same as in normal roots.

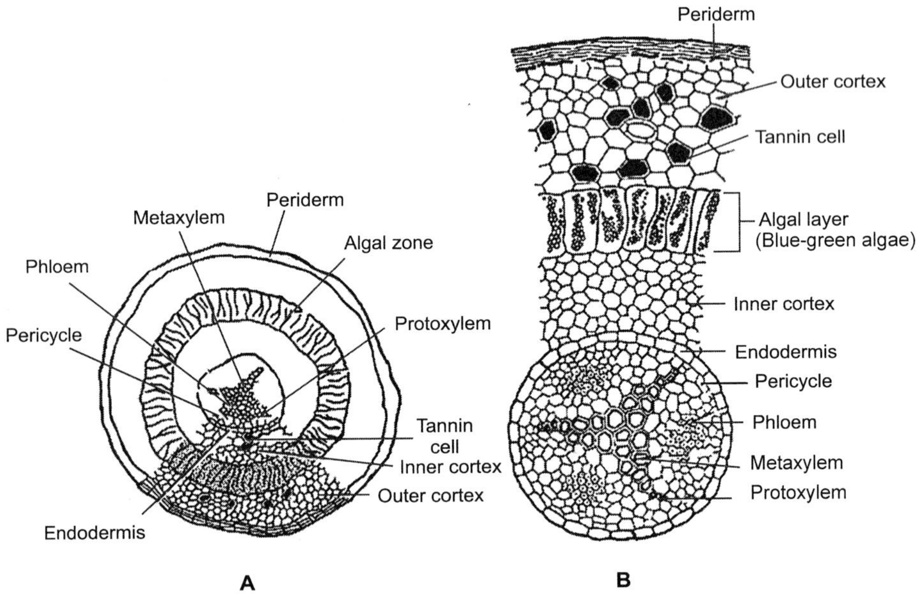

Figure 2.6 A&B: *Cycas* T.S. Coralloid Root.

A - Complete section showing various tissues. B - Sector enlarged to show the discrete algal zone and triarch condition of xylem.

Leaflet

- It is differentiated into a swollen midrib portion and two lateral wings.
- The wings are curved downward at the margins.
- Outermost layer consists of thick-walled epidermis surrounded by a layer of cuticle.
- Upper epidermis is a continuous layer while the continuity of lower epidermis is broken by many sunken stomata.
- Immediatetly below the upper epidermis is present sclerenchymatous hypodermis which is more cells thick in the midrib region than the wing region **(Plate I.4A)**.
- Hypodermis is absent below the lower epidermis, except in the midrib region.
- Mesophyll is differentiated into palisade and spongy parenchyma.
- Palisade mesophyll is present in a continuous layer below the sclerenchymatous hypodermis. Cells of the palisade are radially elongated and filled with chloroplasts.
- Spongy parenchyma is present only in the wings directly above the lower epidermis. The cells are oval, filled with chloroplasts and are loosely arranged with intercellular spaces.
- A single vascular bundle is present in the midrib portion of the leaflet.
- Primary transfusion tissue occurs on either side of the centrally located vascular bundle **(Plate I.4B)**.
- A few layers of transversely elongated cells are present in both the wings (blades) between the palisade and spongy parenchyma. This is secondary transfusion tissue.
- The primary and the secondary transfusion tissues comprise, tracheids and albuminous cells. The tissue is believed to assist water conduction and translocation of minerals.
- The vascular bundle is conjoint, collateral and open. It is diploxylic *i.e.* xylem is of two types *viz.* centripetal xylem and centrifugal xylem. The triangular centripetal xylem is well-developed with endarch protoxylem. Centrifugal xylem is represented by two small groups on either side of protoxylem. Phloem is arc- shaped and remains separated from the cambium. The vascular bundle is surrounded by a bundle sheath.
- The portion of the midrib in between the palisade layer and lower hypodermal region is filled with parenchymatous cells, of which some cells contain calcium oxalate crystals **(Fig. 2.7A,B)**.

Presence of thick cuticle, sunken stomata, hypodermis and transfusion tissue indicate that the leaflet possesses xerophytic characters* *(Plate I.4B).***

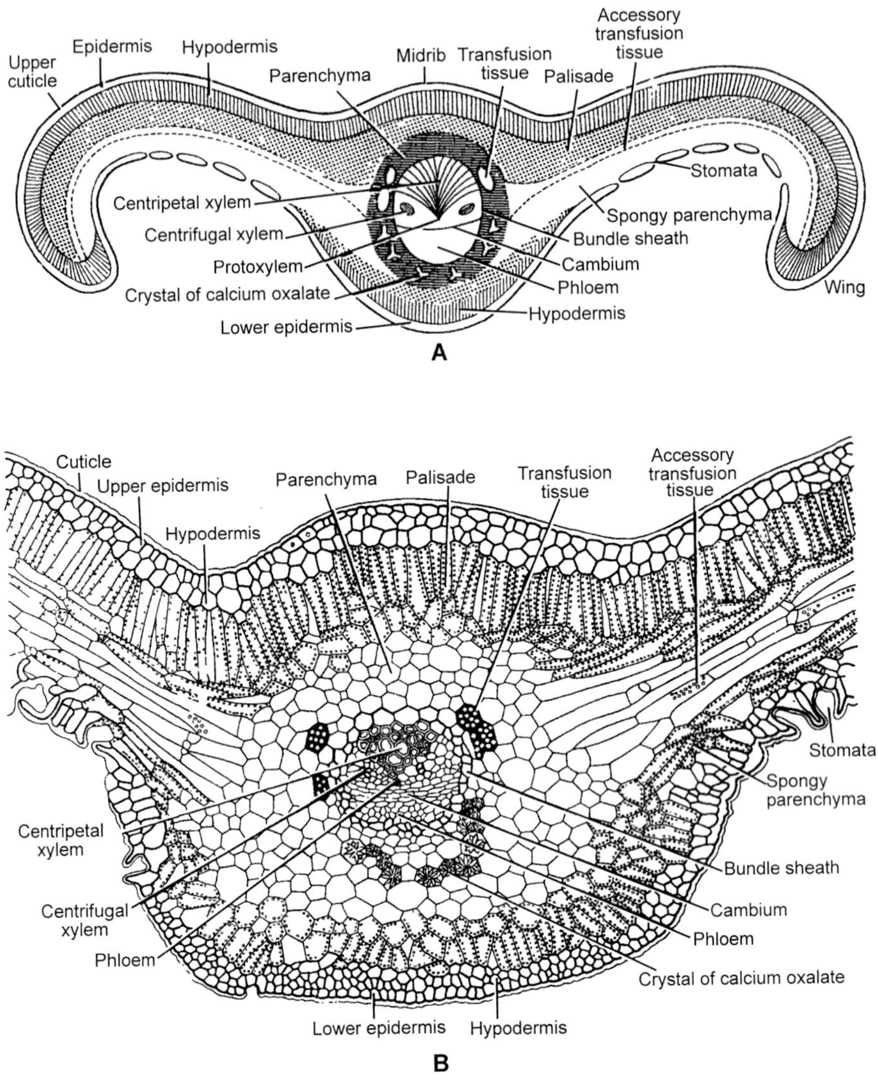

Figure 2.7 A&B: *Cycas* T.S. Leaflet.

A - Outline diagram. B - Cellular details. Structures related to xeric adaptation are clear. Thick cuticle, sclerenchymatous tissue, sunken stomata, and transfusion tissue help conserve water. Mesarch xylem is a primitive character.

Rachis

- The outline of transverse section is characterized by two arms of leaflet bases present one on either side of the rachis that appear like horns **(Fig. 2.8A).**

- Epidermis, the outermost layer of rachis consists of thick-walled heavily cuticularized cells. The entire surface is spanned by irregularly distributed sunken stomata.
- Hypodermis is present below the epidermis. It is differentiated into outer two or three layers of chlorophyll-containing thin-walled cells and inner four to six layers of thick- walled lignified cells of sclerenchyma, poorly-developed on the lateral sides.
- Ground tissue is large, formed of thin-walled parenchymatous cells with mucilaginous canals.
- Vascular bundles are arranged in the shape of an inverted Greek letter Omega (Ω) in the ground tissue **(Plate I.5)**. Towards the tip of rachis the bundles are arranged in C-shaped manner and their number is comparatively less.
- Each conjoint, collateral and open vascular bundle remains surrounded by a bundle sheath **(see Plate 1.5 inset).**
- The xylem in each vascular bundle is present towards inner side. It consists of tracheids and xylem parenchyma while vessels are absent.
- The vascular bundles are diploxylic, *i.e., they* consist of two types of xylem *viz.* centripetal xylem (cpx) and centrifugal xylem (cfx) **(Fig. 2.8B).**
- Phloem, present towards the outer side of the vascular bundle, consists of sieve tubes and phloem parenchyma. Companion cells are absent.

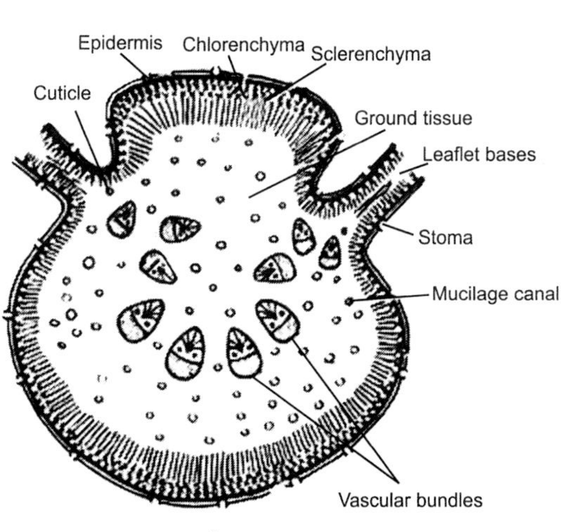

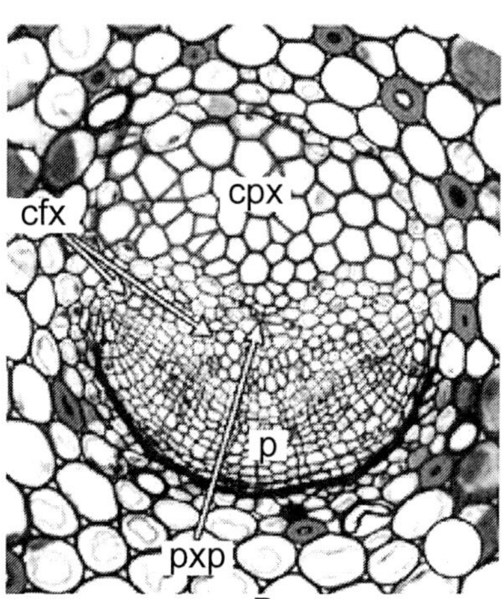

Figure 2.8 A&B: *Cycas* T.S. Rachis.

A - Outline diagram. B - Structural details of vascular bundle are shown. The mesarch xylem is distinct. The phloem (p) is external while xylem is internal. The centripetal xylem (cpx) is opposite protoxylem pole (pxp) while centrifugal xylem (cfx) is on either side of this pole.

Reproduction

Vegetative Reproduction

- Generally occurs by means of bulbils which develop from the axil of the scaly leaves **(Fig. 2.9A).**
- They are more or less oval structures with several scaly leaves arranged spirally and compactly over a dormant stem.
- On detachment from the stem, a bulbil germinates producing many roots towards the lower side and a leaf towards the upper side **(Fig. 2.9B).**
- A bulbil from male plant will develop only into the male plant, while that from the female plant will form only female plant as *Cycas* is strictly a dioecious plant.

A B

Figure 2.9 A&B: *Cycas* Bulbil.

Sexual Reproduction

Cycas is a dioecious plant with separate male and female plants.

Male Strobilus: The Microsporangiate Strobilus

- The male strobilus is probably the largest in the plant kingdom **(Fig. 2.10A)**. It is a compact, somewhat woody, conical or oval structure attached to the tip of the male branch by a small stalk.
- Has a number of leafy sporophylls also known as microsporophylls **(Plate I.6).**
- The development of microsporangia on microsporophylls is acropetal.
- Strobilus also referred to as 'cone' is soft and fleshy when young, becoming hard and woody at maturity.

- The microsporophyll size varies where in ones near apex and at the base are smaller in size while those towards the centre are larger **(Fig. 2.10B).**
- Just prior to shedding of pollen, the strobilus axis elongates rapidly and the microsporophylls get separated from one another.
- Further the microsporophylls below the apex are the first to ripen, dry up and lose turgidity which helps shed their pollen.

Microsporophyll or Stamen

- It is a flattened somewhat triangular structure; narrow below and broadened above with a pointed apophysis **(Fig. 2.10C).**
- On the under surface are present a number of sori with two to six microsporangia or pollen sacs. On an average there are about 1000 pollen sacs per microsporophyll. The microsporangia are arranged radially in sori **(Plate I.7)**.
- The microsporophylls towards the apex and the base are sterile *i.e.,* without microsporangia.

Microsporangium

- Microsporangia are seen as sacs hanging from the under surface of microsporophyll **(Fig. 2.11A; Plate I.8A).**
- The microsprorangia are interspersed with sterile hairs.

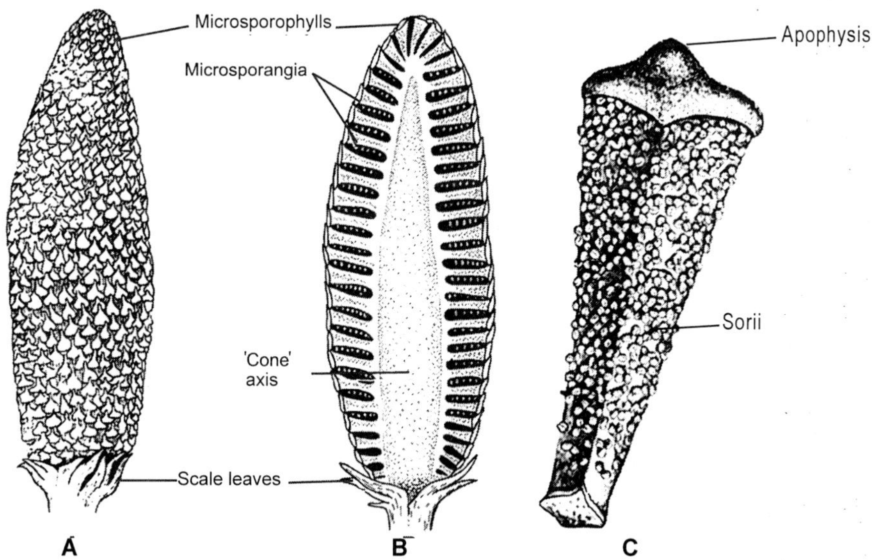

Figure 2.10 A-C: *Cycas.* Male reproductive structures.

A - The male strobilus is large with spirally arranged microsporophylls. B - L.S. of male strobilus to show central axis and microsporangia on the microsporophylls. C-Single microsporophyll is enlarged to show sterile conical portion known as apophysis while fertile portion has sori.

Development of microsporangium is of eusporangiate type with hypodermal sporangium initial, which divides periclinally into an outer sterile primary wall cell. This cell forms three to six layered microsporangial wall. The inner cell forms a primary sporogenous cell that divides through anticlinal and periclinal divisions and forms a mass of sporogenous cells, the outermost of which develops into tapetum **(Fig. 2.11B).** The sporogenous cells get transformed into spore mother cells that undergo meiosis and form haploid microspores. An annulus gets differentiated at the time of dehiscence in the microsporangium which dehisces longitudinally and the microspores or the pollen grains escape.

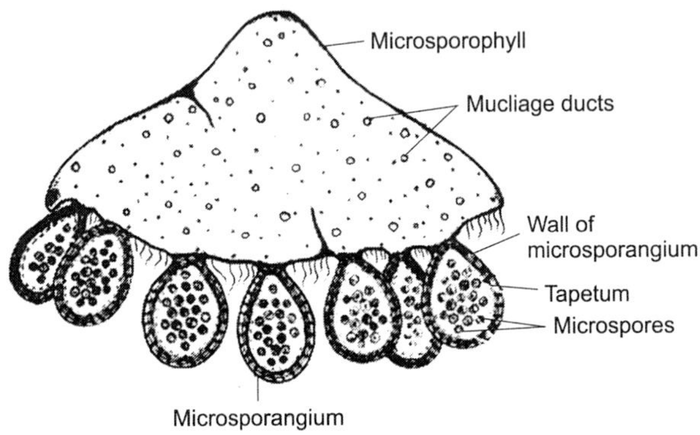

A

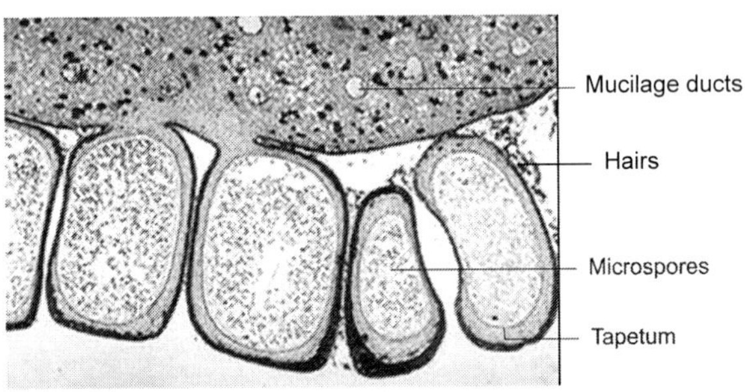

B

Figure 2.11 A&B: *Cycas* V.S. Microsporophyll.

A - Diagrammatic sketch of microsporophyll showing microsporangia on the under surface. The sterile portion of microsporophyll has mucilage canals. B - Magnified view to show cellular details. The microsporangium wall has a distinct tapetum and the locule has microspores. Also seen are hairs interspersed with microsporangia. This is a primitive character.

Arrangement of sori, structure, development, mode of dehiscence of microsporangium and a large number of microspores per microsporangium are features in which Cycas resembles Angiopteris and shares the characters therefore, with pteridophytes* *(Plate I.8B)****.*

Female Strobilus: Ovulate 'cone'

- The largest among plant kingdom.
- It is not a proper 'cone' but simply a rosette of megasporophylls arising spirally in acropetal manner.
- The megasporophylls are loosely arranged like leaves and a complete crown of megasporophylls is formed once a year **(Plate I.9).**
- The strobilus is erect first but droops at maturity.

Megasporophyll or Carpel

- Has three parts - proximal or the petiole, middle stalk-like fertile portion with ovules and the distal portion which is lobed and serrated. The megasporophyll bears hair that are shed at maturity and the upper portion is chlorophyllous when young.

**Peculiarities*

It resembles a reduced foliage leaf with three parts, lower - which is petiole like, middle - fertile region with lateral ovules that replaces pinnae on this part and the upper - sterile region whose morphology is a taxonomic feature ***(Fig. 2.12A-C).*** *Presence of hairs is noteworthy. Megasporophylls are therefore, modified foliage leaves.*

In *C. rumphii,* the upper sterile portion of the megasporophyll is pinnate though reduced, and tapers to a point.

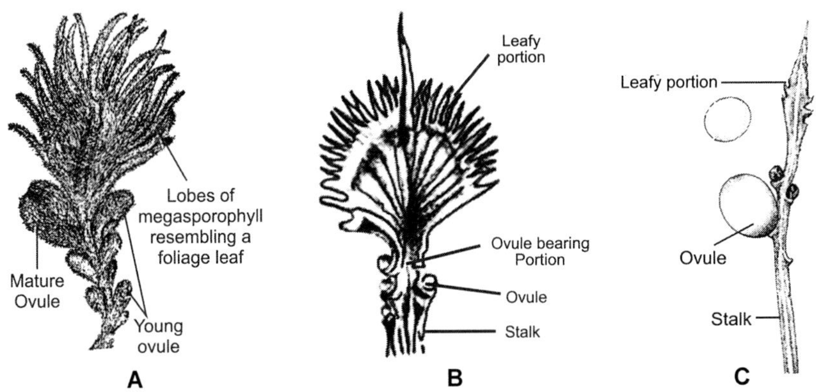

Figure 2.12 A-C: *Cycas.* Female reproductive structures – megasporophylls.

A - *Cycas revoluta.* B - *Cycas siemensis.* C - *Cycas rumphii.* Note variously dissected sterile, distal ends which is a taxonomic feature.

Megasporangium or Ovule

- It is orthotropous, unitegmic, short-stalked, large and showy.
- It arises hypodermally as a mass of meristematic cells on megasporophyll; the mass undergoes mitotic divisions and results in a massive nucellus.

- From the base of the nucellus arises single integument, which forms a micropylar tube towards the apex.
- Later at the time of pollination, some cells of nucellus break down forming a pollen chamber and pollen grains germinate here **(Plate I.10).**
- Integument is thick and in mature seed is divisible into outer fleshy, middle stony and inner fleshy layers. It is traversed by two vascular bundles, one to the outer layer and second to the inner /middle layer.

Gametophytes

Male Gametophyte: Pollen Grains

- Pollen grain is boat-shaped with a longitudinal slit.
- The pollen wall has intine and exine, and a centrally placed nucleus which shifts to one side and divides into a small vegetative or the prothallial cell and a large antheridial cell.
- At the time of dehiscence, pollen is a three celled structure with a vegetative cell, and a generative and tube cell (formed by mitotic division of antheridial cell) **(Fig. 2.13A-C).** Some workers believe that generative cell first divides into a body cell and a stalk cell where the former divides into two spermatozoids.
- After pollination, a pollen tube is formed from the tube cell and pollen grain lies embedded in female tissues. The generative cell that lies next to the prothallial cell may be regarded as the primordial germ cell which is the progenitor of two male nuclei.

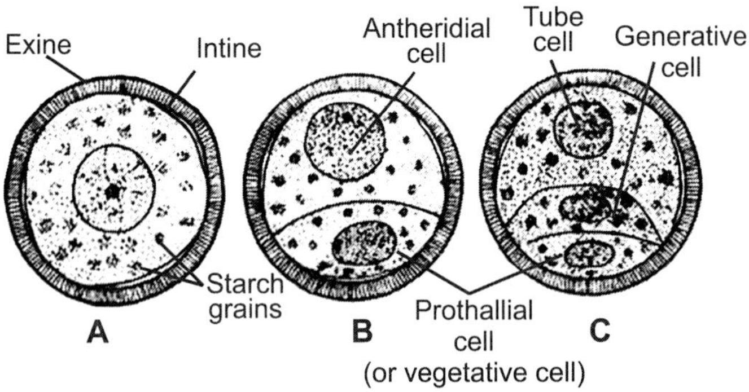

Figure 2.13 A-C: *Cycas*. Male gametophyte development.

A - Uninucleate haploid pollen grain. B - Pollen grain with a single prothallial cell and an antheridial cell. C - Three celled stage.

Megaspore and Female Gametophyte

- All cells of nucellus are alike in the beginning but later a megaspore mother cell (diploid) gets differentiated from a deep seated nucellar cell which possesses a large and dense nucleus.

- Reduction division of the mother cell results in a linear tetrad of megaspores (haploid), out of which only the lower most is functional.
- The haploid nucleus of the functional megaspore enlarges and mitotic divisions ensue. Free nuclear divisions lead to formation of about 1000 nuclei.
- A vacuole appears in the centre and the free nuclei get arranged in the periphery.
- Later cell walls are laid down in a centripetal fashion; from outside towards inside till the vacuole gets obliterated and the cellular female gametophyte or female prothallus or endosperm is formed **(Fig. 2.14 A-C)**.
- The female gametophyte has small cells above and large cells below; latter store food for the embryo.
- At the same time, cells of the inner layer of the nucellus form an endosperm jacket or spongy tissue (named by Strasburger).
- As the female gametophyte becomes cellular, sugar is the principal food content, but later starch dominates. Some cells however, contain tannins.
- A few cells of nucellus at the micropyle end degenerate forming a pollen chamber.

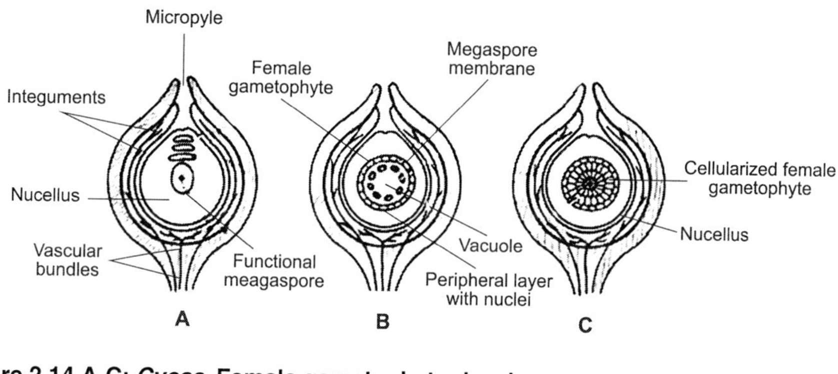

Figure 2.14 A-C: *Cycas*. Female gametophyte development.
It develops from a linear tetrad of megaspores to form a cellular gametophyte.

Pollination

It was once thought that pollination of cycads occurred randomly by wind blowing the pollen from male strobili to nearby female strobili. The large male strobili are well adapted for this kind of pollination mechanism and shed pollen in large number in the form of 'pollen cloud'. For example, *C. circinalis* sheds 100 cubic centimetres of pollen. However, in most cycads, the pollen is less in amount when the strobili are small and also the scales of female strobili are so placed (often drooping) that pollen cannot reach egg through wind agency. There are now reports that insects can bring about pollination **(See Box 2.1)**. Evidences suggest that if cycad is removed from its native area which also has its insect pollinator, the plant will remain sterile because of the absence of insect pollinator in the new habitat (Norstog *et al.*, 1986).

Scientists at University of Utah discovered a strange method of reproduction in cycads: the plants heat up and emit a toxic odour to drive pollen-covered insects out of male cycad structures, and then use a milder odour to draw the bugs into female strobili so the plants are pollinated (www.science20.com › Applied Physics › News Articles).

Box 2.1

Pollination in Cycadaceae remains controversial. Experiments suggest that pollination in *Cyas circinalis* may have two phases: the transport of wind borne pollen grains to megasporophylls and then subsequent transport of adhering pollen to ovules by water and/or wind. In *Cycas panzhihuaensis*, Wang *et al.* (1997) suggested that after pollen grains are first transported by wind from male to female strobili, they may be carried forward to the ovules in the female by insects. In *C. revoluta, C. rumphii* and *C. thouarsii* both male and female strobili emit a strong odour and a temperature increase mechanism for insect pollination.

**There is not doubt that C. revoluta can be pollinated by wind. However, because the amount of the airborne pollen decreases markedly beyond a 2 m radius from male strobili, wind pollination is limited unless female strobili are located near the male.*

Post Pollination Changes

- A week after pollination, starch grains disappear from the male gametophyte cytoplasm, a central vacuole appears and the nucleus moves to a side.
- A three week old pollen grain is a three celled structure with vegetative (prothallial) cell, tube cell and a generative cell.
- Post pollination, tube cell forms a pollen tube, which elongates piercing through the exine to form a long and branched pollen tube and the nucleus moves to the centre.
- Pollen tube acts as a haustorium which gradually grows into the nucellar tissue digesting it at the micropylar end. It is also a carrier of male gametes.
- The generative cell divides into a stalk and a body cell **(Fig. 2.15A).**
- Later two motile spermatozoids arise from the body cell **(Fig. 2.15B).**

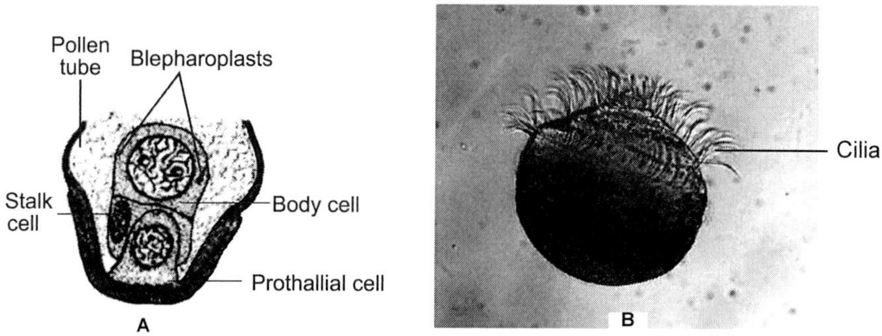

Figure 2.15 A&B: *Cycas*. A - Tip of the pollen tube with body cell and prothallial cell which pierces through the stalk cell. This is a post pollination development. B - The ciliated spermatozoid is a primitive character.

- On the female side, the nucellar beak closes and hardens shutting the germinating pollen in the pollen chamber.
- A few cells of female gametophyte at the micropylar end become large and are known as archegonial initials. The number of these may range from two to eight but three is common.
- Each archegonium has two neck cells, a short - lived ventral canal cell (VCC) represented by nucleus only, and an egg cell **(Fig. 2.16A).**
- Absence of neck canal cells and a short-lived VCC is an evolutionary feature.
- During development of the archegonia, surrounding cells of the nucellus in the micropylar region outgrow forming a shallow basin or the archegonial chamber. This gets filled with liquid during fertilization.
- At the time of fertilization, nucellar tissue between pollen chamber and archegonial chamber disorganizes forming a continuous cavity **(Fig. 2.16B).**

**After pollen is delivered to ovules via wind or insects, there are complex interactions between pollen and ovules prior to fertilization in seed plants. These interactions are integral in the functional reproductive biology of seed plants. Among extant gymnosperms there are diverse pollen delivery mechanisms that bring pollen into the ovule, where it germinates. Pollination mechanisms can be classified into types according to presence and absence of pollination drops, and by pollen grain and/ or ovular modifications. Although most pollination mechanisms of gymnosperms can be grouped into four pollen capture mechanisms (PCMs) that describe pollen-ovule interactions, there are a few gymnosperms in which ovules neither capture the pollen directly, nor are they the site of germination. These mechanisms as extra-ovular capture and germination (ECGs). In most gymnosperm genera, pollen grains germinate in pollination drops-liquid secretions containing minerals, carbohydrates and proteins, including active enzymes. Once germinated, pollen interacts primarily with the nucellus.*

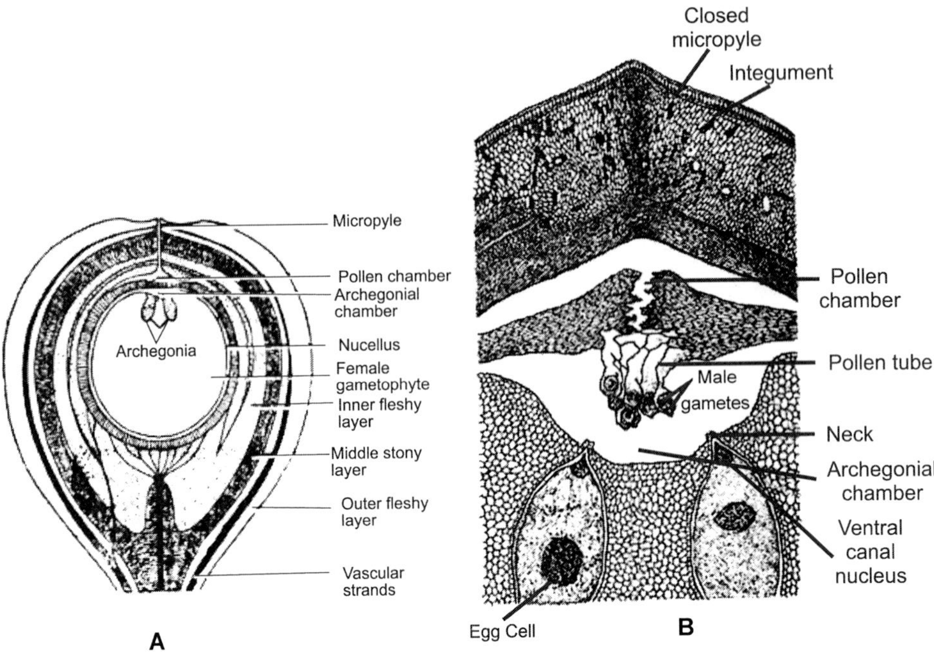

Figure 2.16 A&B: *Cycas*. Female gametophyte and post pollination changes.

A - Formation of archegonia and pollen chamber is significant. B - Pollen grains are seen to germinate in the pollen chamber and pollen tubes travel into archegonial chamber. The micropyle is closed and the archegonia harbor egg cell.

Pollination to fertilization is a long gap (about three to four months. In some species it may be six months).

Pollination occurs in June or July and is shortly followed by the formation of the pollen tube. From this stage till the time when spermatozoids are produced, about two months pass. Fertilization occurs in the month of September or October.

- The pollen tube pierces through the nucellus and hangs freely in the cavity which is partly archegonial and partly a pollen chamber.
- The stalk cell enlarges considerably, is filled with starch and nutritive material and gets penetrated by the prothallial cell.
- The body cell enlarges and has a prominent nucleus with two blepharoplasts; it gives

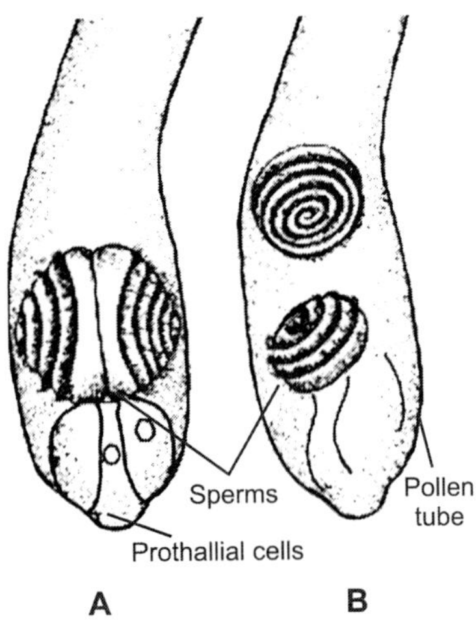

Figure 2.17 A&B: *Cycas* Male Gamete.

A - Formation of gametes from the body cell while the prothallial cell is vestigial. B - The spermatozoids are large and ciliated structures.

rise to two spermatozoids with cilia **(Fig. 2.17A&B).**

- Spermatozoids which were discovered by Hirase are visible to the naked eye, are top shaped, and naked with a ciliated anterior.
- Spermatozoids swim in the cavity of the body cell for about 30 min, separate and pass into the pollen tube tip. The cilia begin to move while the spermatozoids are still attached to stalk cell. This movement is accompanied by pulsating and amoeboid movement of cytoplasm and the nucleus of male gametes.
- Before spermatozoids are discharged, pollen tip becomes more and more turgid and then ruptures. One of the spermatozoids escapes in two or three seconds but the other may take about half a minute.

Fertilization

Fertilization has evolutionary significance as it exhibits both siphonogamy (*i.e.* formation of pollen tube) and zooidogamy (*i.e.* participation of ciliated male gametes).

In cycads, spermatozoids are released from pollen tubes and swim in fluid toward the archegonia. The source of this fluid was examined in culture using *Cycas revoluta* ovules. Dissected female gametophytes just before fertilization produced copious fluid on their upper surface. The fluid first appeared around the archegonial chamber and then on the inside of the archegonial chamber. When this fluid was applied to dry turgid pollen tubes, they discharged spermatozoids 12 h later. To bring about fertilization, a spermatozoid makes its way through the fluid which has accumulated around the egg cell, bores into the egg cell and loses its cilia and cytoplasm, after which its nucleus moves towards the egg nucleus sinking into a depression on the peripheral end of the egg (http://archive.org/stream/jtor-2454374/2454374_djvu.tx0.2). The archegonial neck appears as two semi-spherical swellings, and many globose particles appear on the top of the archegonial neck cells when the fluid is present. The contents of pollen tubes, spermatozoids and surrounding liquid intermingle with the secreted fluid. The female gametophyte differs in ultrastructure during the stages before and after fluid secretion, latter showing changes suggestive of fluid secretion from the female gametophyte (Takaso *et al.,* 2012).

- Before fertilization, the ciliated membrane of male gametes has to be given off and it slips to release the male nucleus which fuses with the egg to form zygote.
- According to Chamberlain (1935), just prior to fertilization, the portion of the egg cell protrudes into the archegonial chamber. This portion which is vacuolate, strikes with spermatozoids so violently that ciliated membrane gets torn off and remains on egg top. Nuclear fusion occurs subsequently between male and female gamete.
- According to De Silva and Tambiah (1952), when turgidity of neck cells

is lowered, spermatozoids are drawn inside due to chemotaxis. Though several spermatozoids enter but only one fuses with the egg, rest degenerate.

- The fertilization is described as Zooidogamy (zoids + siphonogamy), observed by IKeno (1896). Also seen in relic fossils.

Embryogeny

- Zygote does not undergo a period of rest but develops to form a proembryo.
- The large 2n nucleus undergoes free nuclear divisions and about 256 nuclei are formed **(Fig. 2.18A).** Central vacuole is formed and 256 nuclei are pushed towards periphery.
- Cell walls later appear and vacuole gets obliterated leading to a proembryo formation.
- The proembryo differentiates into three regions; haustorial in contact with nutritive tissue, suspensor with elongated cells and embryo with meristematic cells **(Fig. 2.18B).**
- Suspensor with elongated cells is torturous and is spirally coiled; it pushes the embryo into the endosperm, the nutritive tissue **(Fig. 2.18C).**
- Though all archegonia may get fertilized and suspensors may twist to form a compound structure with polyembryos, but only one embryo reaches maturity **(Fig. 2.18D, E).**
- Embryo is a straight structure, has a radicle, plumule and cotyledons.

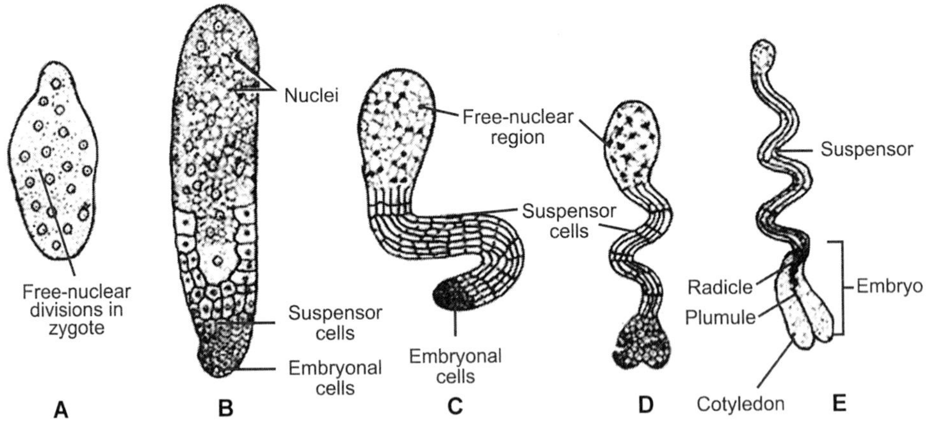

Figure 2.18 A-E: *Cycas.* Stages in formation of proembryo.

As the development progresses, suspensor differentiates and later elongates carrying embryonal cells at the tip.

Seed

- The seeds at maturity are colourful **(Plate I.11).**
- Ovular integuments form the seed coat, outer and inner layers are fleshy while middle layer is stony **(Plate I.12A&B)**. At maturity inner layer which is fleshy is also consumed, coleorrhiza and radical lie next to suspensor, while endosperm fills the seed. Cotyledons are two which may or may not be equal or only one is developed while second remains suppressed **(Fig. 2.19A).**
- As testa (outer layer) is coloured, sweet and with odour, seed is dispersed by birds. The seed has no rest period as it is viable for a few months only. It germinates quickly and the radicle gives rise to the root (**Fig. 2.19B; Plate I.12B).** The cotyledons help in absorbing the nutrients and are, therefore haustorial in nature. Later they shrivel and remain attached to parent plant for a year or so. The first few leaves are the scale leaves.

**The seeds of Cycas are well known for being acutely toxic and carcinogenic. Researches have shown that seeds have compounds which make them neurotoxic (Spencer et al., 1987). However, they are processed, made nontoxic and are then consumed by humans.*

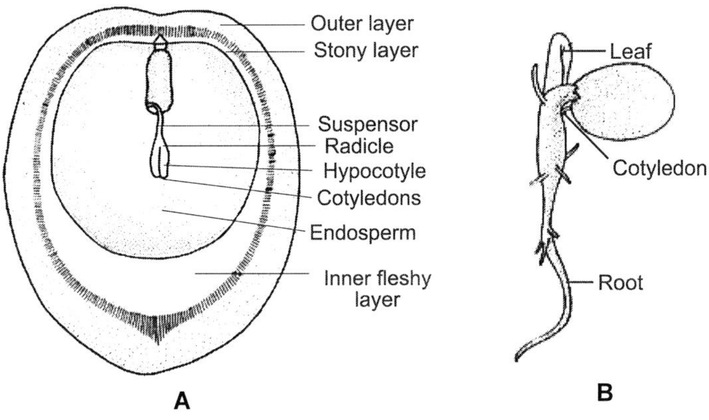

Figure 2.19 A&B: *Cycas* Seed.

A - L.S. seed to show well differentiated coat and a massive endosperm. B - Germinated seed is seen with first leaf and a root.

The life cycle of *Cycas* is completed in four years:

- ***First year***: March to July, male and female structures attain maturity; with pollination, pollen tube formation, and generative cell dividing into body cell and stalk cell.
- ***Second year***: March to December, pollen tube grows and blepharoplasts produce cilia.

- ***Third year***: June to November spermatozoids are formed, fertilization ensues and embryo gets organized.
- ***Fourth year***: January (end) to March is marked by seed dispersal, organization of embryonal axis and seed germination.

Salient Features of Sexual Reproduction

The archegonial development in *Cycas* has three distinct periods:

First, a period, in which the archegonial cells are differentiated is followed by the second period of growth during which the egg cell attains a relatively enormous size with a nucleus approximately 75-120 microns in diameter. The nuclei of the surrounding cells add to the nourishment to the central cell when they pass through pores in the cell walls. The third phase is the maturation period in which the central cell prepares for the act of fertilization and divides unequally to produce a small canal cell and an oval, ventral egg cell.

Do You Know?

Total of 4,200 expressed sequence tags (ESTs) were created from *Cycas rumphii* and clustered into 2,458 contigs of which 1,764 had low-stringency BLAST similarity to other plant genes. Among those cycad contigs with similarity to plant genes, 1,718 cycad 'hits' are to angiosperms, 1,310 match genes in gymnosperms and 734 match lower (non seed) plants. Forty six contigs were found that matched only genes in lower plants and gymnosperms. Upon obtaining the complete sequence from the clones of 37/46 contigs, 14 still matched only gymnosperms (Brenner *et al.*, 2003).

Test Your Knowledge

Q 1. *Cycas* is a 'living fossil'. Comment.

Q 2. Discuss the secondary growth in stem of *Cycas*.

Q 3. Why is it erroneous to call female strobilus, a cone in *Cycas*?

Q 4. Give our new understanding of pollination mechanism in *Cycas*.

Q 5. Write a short note on:

Spermatozoids of *Cycas*, pollen chamber, megasporophyll, polyembryony, manoxylic wood.

Chapter 3

Subclass Ginkgoidae

Relationships of the extant genus *Ginkgo*, with other gymnosperms have been long debated. It is apparently not related to any other living plant. Based on certain reproductive features such as flagellated male gametes seen in both, *Ginkgo biloba* was thought to be closely related to members of Cycadales (Chaw *et al.*, 1997) and therefore, Ginkgoidae has been suggested to be a sister group to the cycadophytes **(Fig. 3.1).** In another interpretation, hypothesis was advanced

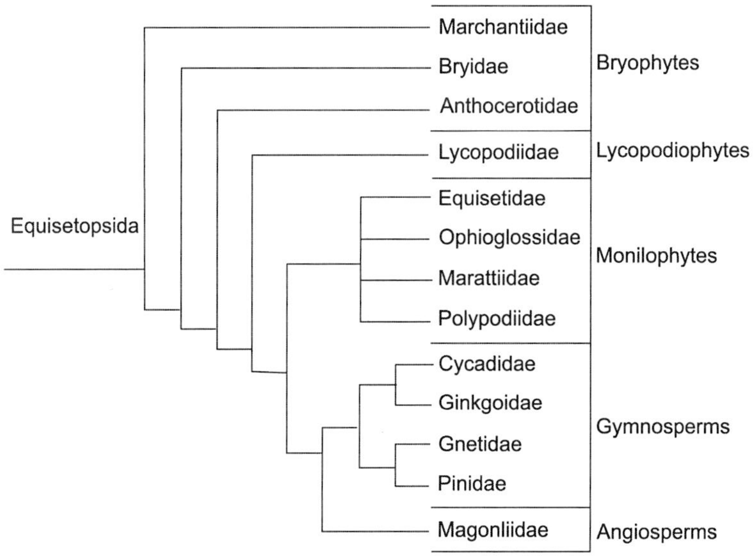

Figure 3.1: *Ginkgo*. Phylogenetic relationship of the genus with other subclasses (after Christenhusz *et al.*, 2011).

that Ginkgoidae diverged from some Carboniferous pteridosperm group and evolved separately since then (Meyen, 1982, 1984). Doyle and Donoghue (1986) related ginkgos to conifers and cordaites, while Bowe *et al.* (2000) related the group to a clade including conifers and gnetophytes. Based on a cladistic analysis using cordaites and early conifers as the out group, Zhou (1991) suggested Ginkgoales to be a monophyletic group that included extinct *Trichopitys, Karkenia, Toretzia, Umaltolepis, Yimaia*, and extant *Ginkgo*. However, a recent comparative morphological and morphogenetic study of male sporangiophores (Mundry and Stützel, 2004) suggests a relationship of *G. biloba* with Coniferales, Gnetales, and Cordaitales, and rejects a close relationship with groups that have pinnately organized sporophylls, such as Cycadales, as assumed by some molecular studies.

Is *Ginkgo* a Living Fossil ?

A group of gymnosperms of great interest to paleobotanist with three genera - *Ginkgoites, Baiera* and *Sphenobaiera* and extant *Ginkgo*, originated over 250 million years ago. The origin dates back to Permian **(Box 3.1).**

During the middle Jurassic, ginkgoales experienced a great increase in species and reached their maximum diversity during Cretaceous. During Paleocene (Late Cretaceous and Tertiary periods), Ginkgoalean diversity reduced to a single polymorphic species referred to as *Ginkgo adiantoides.* This species is almost indistinguishable from the extant *Ginkgo biloba.* During the Quaternary, *Ginkgo* became restricted to small areas in eastern China. The only known wild population of *Ginkgo biloba* is in a temperate moist forest of Zhejiang. *Ginkgo biloba* is deciduous and the abundant occurrence of leaves in certain geological horizons indicates that at least some extinct ginkgos were also deciduous.

Previous fossils revealed that *Ginkgo* species have remained unchanged for the past 5 million years and that similar trees were alive 170 million years ago during the Jurassic period.

Fossil from 121 million years old Yixian rock formation in China has proved that *Ginkgo* morphology has changed little over 100 million years. In the Yixian formation a single species of *Ginkgo* was reported by Zhou and Zheng (2003). It is intermediate in the ovulate structure between modern *Ginkgo* and Jurassic ginkgos with many ovules. The leaves of the Jurassic plants were divided with several lobes. Some 50 million years later the lobes had joined up into small fans of today's specimen. The way that seeds formed has also changed. Today's *Ginkgo* makes a few seeds, only one of which reaches full maturity in a single stalk. Jurassic trees possess sprays of stalks each sporting a single seed.

Thus, today's *Ginkgo* is a direct descendent of forebeares that co-existed with dinosaurs (www.nature.com, news June 2003).

Box 3.1

Ginkgo has a long fossil record and is well known as having pre-eminent claim to be described as a living fossil (Seward, 1911, 1919). Living *Ginkgo* is called 'Baiguo' (white nut), Yajiao (duck foot), gongsunshu (grandfather grandson tree). However, the name *Ginkgo* originates from 1712 when the traveller Kaempfer proposed this plant. Linnaeus adopted this and added species name 'biloba' to denote the bisection of the wedge-shaped lamina of the leaf into two divergent segments. There is no direct fossil evidence and no unanimously accepted theory about the origin of ginkgoales, although most believe that they originated from progymnosperms related either to pteridosperms and cycads or to cordaites and conifers (Kenrick and Crane, 1997). It is believed that *'Ginkgo'* is an oddity in nature because it is a single species with no known living relatives; a living fossil that has been essentially unchanged for more than 200 my and an inspiring example of how humans can help a species survive (in interview by Roger Cohn, 2013).

Order Ginkgoales

Comprises large trees characterized by long shoots and short or dwarf shoots. They have wood with small pith, and proliferate helical branching with leaves borne in a helical arrangement. Leaves on long shoots are typically widely spaced while those on the short shoots are clustered around the apical part. Leaves are unique, fan-shaped with dichotomously branching veins, bilobed or sometimes strongly dissected. The only extant species, *Ginkgo biloba*, is dioecious with female and male trees. The ovules and male strobilus are terminal and borne on the short shoots which in case of former are called ovulate axis. Two ovules are borne in small cups on a ovulate axis. However, only one reaches maturity forming large seeds with outer fleshy layer.

Ginkgo

Distribution

The name *Ginkgo* comes from Chinese word yin-kuo (silver fruit). Japanese call *G. biloba*, I-cho meaning "tree with leaves like a duck's foot". *Ginkgo,* is classified under subclass Ginkgoidae comprising order Ginkgoales and family Ginkgoaceae.

The largest and oldest *Ginkgo* in the world may be the so-called 'Grand Ginkgo King' (Xiang, 2009). This stands about 30 meters tall with a ground level diameter of 460 cm. The internal tissues with their growth rings of the oldest part of the multi-generational trunk are not clear; but the age has been estimated as 4,000-4,500 years. It is found in the hamlet of Li Jiawan in China. Ginkgos are now used as ornamental plants; in China and Japan the seed kernels are a foodstuff; and it is claimed that the leaves have medicinal properties (*Ginkgo* - *Ginkgo biloba* - Details - Encyclopedia of Lifeeol.org/pages/1156278/details).

Four ginkgo trees that survived the 1945 atomic bomb explosion in Hiroshima, even though they were less than a mile from the point of impact are known to have recovered well and can be seen there even today. In China and Japan it is grown as a sacred tree in temple gardens. It is cultivated in the United States as a shade tree. It is also successfully cultivated in some gardens of Europe, America and India.

Do You Know?

Ginkgo biloba is rated as Endangered (EN) according to IUCN Red List criteria and has received a EDGE score of 4.89, which places it in position 2 on the EDGE Gymnosperm list. It has Evolutionary Distinctiveness of 315, ranked 1 (https://www.edgeofexistence.org).

Sporophyte

It is a large tree normally reaching a height of 20–35 m, with some specimens in China being over 50 m (164 feet). The tree has an angular crown and long somewhat erratic branches, and is usually deep rooted. It features distinctive two-lobed, somewhat leathery, fan-shaped, rich green leaves with diverging (almost parallel) veins **(Fig. 3.2)**. Leaves turn bright yellow in fall. *Ginkgo* trees are commonly called maidenhair trees in reference to the resemblance of their fan-shaped leaves to maidenhair fern leaflets (pinnae). *Ginkgo* is dioecious (separate male and female trees). It is resistant to wind and snow damage. A combination of insect and disease resistant wood and the ability to form aerial roots and sprouts makes it a long-lived tree with some trees claimed to be more than 2,500 years old. *Ginkgo* is a unique tree with no close living relatives (Bego, 2011).

Ginkgo is remarkably tolerant of difficult conditions, including pressures of urban pollution. Because of this, it is widely planted in cities.

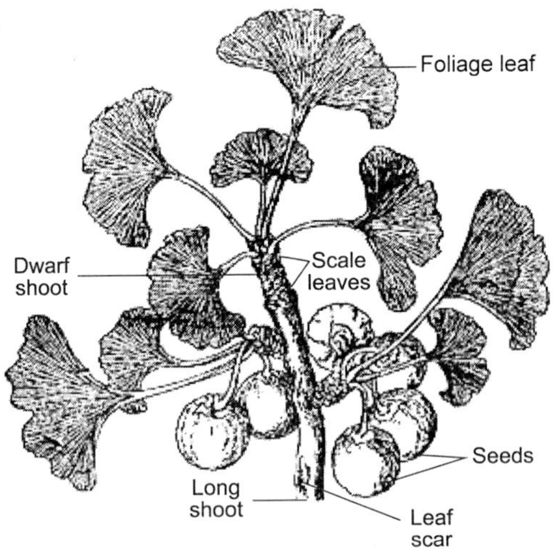

Figure 3.2: *Ginkgo* Sporophyte.

Portion of a reproductive axis. The dimorphic branches and leaves are distinct. Seeds are borne by the spur shoots or the ovulate axis.

Morphology

Stem

- The trunk is strongly ex-current and the outline steeply pyramidal. As the tree gets about 80 years old canopy begins to broaden, so that the outline appears rounded at the top.
- There are two kinds of branches - the long shoots and the short spur or dwarf shoots **(Fig. 3.3A&B).**
- Long branches grow in length with regularly spaced leaves. From axils of these leaves, 'spur shoots' (short shoots) develop on second year growth **(Plate II.1).**
- Spur shoots have short internodes (grow only one or two centimeters in several years) and their leaves are usually not lobed. Because of the short internodes, leaves appear to be clustered at the tips.
- Reproductive structures are formed only on these spurs **(Fig 3.3A)**.

**The girth of the trunk of older trees may become large because of secondary growth. The tree usually loses its main trunk leader as several vertical trunks (basal chichi) that keep reaching great heights are formed. These are also called lignotubers which give rise to new plants.*

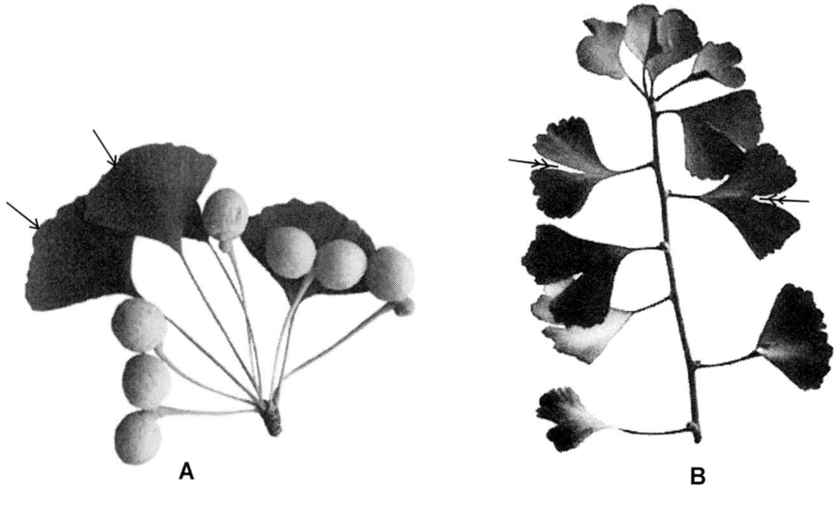

Figure 3.3 A&B: ***Ginkgo*** **Dimorphic Shoots.**

A - Spur shoots with leaf cluster. The lobes of these leaves are indistinct (arrows).
B - Long shoot is seen with regularly spaced leaves. The leaves are distinctly lobed (double arrows).

Root

- *Ginkgo* has relatively primitive and simple root system.
- Roots go more down instead of out and close to the surface.
- Unlike other trees, *Ginkgo* can produce a kind of aerial roots. These structures, called chichi, are produced along the trunk and branches **(Fig. 3.4).** They grow down into the ground, where they form roots, as well as upward, where they sprout leaves. Chichi are usually found on old trees, but can be found on young trees in response to environmental stress or damage.

Figure 3.4: *Ginkgo* aerial roots (arrows) also known as lignotubers or chichi roots.

Leaf

- The leaves on spur with their symmetrical dichotomous venation resemble that of *Adiantum*, the maiden hair fern.
- The dichotomous venation however, does not anastomose.
- One or more (three or four) leaves grow alternately on the branches of unlimited growth in one node.
- The leaf is small in the beginning, gradually becoming three or four inches wide **(Fig. 3.5)** and has a long petiole (Bego, 2011). Those on first year long shoots, and of seedlings, are deeply lobed **(Plate II.2).** Besides the two deep primary lobes, leaves may have two or three secondary lobes on each side, so that they approach the deeply and narrowly lobed leaves of the extinct *Baiera*.

Figure 3.5: *Ginkgo* leaf shows a long petiole, dichotomous venation, wavy margin and two primary lobes (arrows).

- The leaves on the spurs usually have only a wavy margin, are without deep lobes and are closely placed. In the petiole there are two strands, each, by repeated forking, forming the venation of its side of the leaf (Chamberlain, 1935).
- The leaves turn a clear golden yellow in the fall season. *Ginkgo* typically drop its leaves in a very short period of time. Some trees drop all their leaves in a single day!

Anatomy

Stem

- In transverse section, the outline of the long shoot and the spur is very different.
- The long shoot has a comparatively small pith and cortex, a well developed xylem, and a few mucilage cavities.
- However, in spur shoots, xylem is poorly developed but the amount of pith and cortex is more. Mucilage cavities are abundant.
- In both, tannin cells and calcium oxalate crystals are abundantly and widely distributed.
- The vascular cylinder of the stem is an endarch siphonostele. The protoxylem consists exclusively of thickened tracheids, which are much more abundant in the spur shoots than in the long shoots.
- In the long shoots, in the radial direction, there are one to five spiral elements; while in the spur shoots there may be as many as ten.
- The tracheids of the secondary wood have one or two rows of bordered pits on the radial walls arranged in opposite manner.
- Bars of Sanio (thickened portion below the pits) are easily demonstrated in the secondary wood, but do not occur in the primary wood.
- The medullary rays generally are one cell in height and width. Rays one cell wide and two cells high predominate in the trunk and in the long shoots, but some rays are three, four, or even five cells in height. In the spur shoots the rays although only one cell wide, range from one to sixteen cells in height, while rays three to six cells high are common.

Root

- Outermost is phelloderm followed by cortex.
- The root is diarch at young stage and becomes tetrach or hexarch at maturity.
- Endodermis shows conspicuous thickenings on its radial walls. Secondary growth occurs in usual manner.

- Some of the cells of xylem parenchyma have crystals. Xylem rays vary in height from one to seven cells and often show crystals. Thick-walled fibres are abundant in phloem.

Leaf

- A unistratified epidermis of papillomatous squared cells, with a strong cuticle and sunken stomata comprises the epidermal system.
- A discontinuous hypodermis composed of one to three layers of thick-walled cells, alternating with thin-walled cells is present and is followed by pluristratified chlorenchyma of isodiametric cells (containing few calcium oxalate druses) with small intercellular spaces. It exhibits generally four schizogenic cavities symmetrically arranged, three toward the flat face and the fourth toward the middle part of the convex face.
- Each circular-shaped cavity is composed of eight to twelve mucilage secreting epithelium cells.
- The parenchyma sheath of the conducting system presents scanty idioblasts with 34-55 μm calcium oxalate druses in the cell lumen.
- Intercalated among the vascular bundles, schizogenic cavities are present.
- The phloem possesses uniseriate small polyhedric calcium oxalate crystals (15-23 μm), and scanty albuminous cells.
- The xylem is composed of narrow and wide tracheids with thick lignified walls, radially arranged and uni-biseriate parenchyma rays.
- In the lamina region sunken elliptical stomata measuring 53 μm x 33 μm are spread without specific orientation.
- The mesophyll possesses a poorly developed palisade chlorenchyma composed of a continuous layer of generally lobulate cells diminishing in size towards the margin.
- The spongy chlorenchyma of irregular cells with intercellular spaces is localized close to the abaxial face.
- Mucilaginous idioblasts and isolated druse-shaped calcium oxalate crystals measuring 20 μm to 31 μm are present in the mesophyll.
- There are numerous collateral open vascular bundles limited by uni-stratified endodermis with thin-walled large cells.
- The transfusion tissue is represented by reticulate tracheids localised close to the xylem.

Reproduction

Vegetative Reproduction

Ginkgo reproduces vegetatively through peg-like structures – aerial chichi that give rise to roots and leafy branches when they grow into the ground because of embedded vegetative buds.

Sexual Reproduction

Ginkgo is dioecious though male and female plants are difficult to differentiate when young. Male tree usually has a slim column and is slightly longer, the female tree has a wider crown and a more spread out form.

Reproductive bodies are most primitive among living seed plants except some Cycadales.

Male Strobilus: Microsporangiate Strobilus

- Male or microsporangiate fructifications develop in catkin-like clusters on the dwarf or spur shoots of male plants **(Fig. 3.6A; Plate II.3).**
- Each male strobilus contains several microsporophylls arranged loosely on a central axis.
- Each microsporophyll has a long stalk terminating into a hump or knob.
- It contains two pendant microsporangia. A mucilage duct is present in the knob **(Fig. 3.6B).** Rarely more than two microsporangia are present in a microsporophyll.

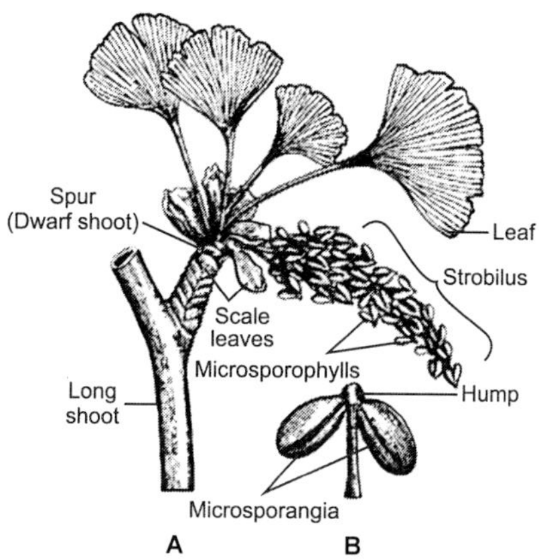

Figure 3.6. A&B: *Ginkgo.* Male reproductive structures.

A - Catkin-like male fructifications on spur shoots. B - Single microsporophyll to show two pendant microsporangia and a hump.

Microsporangium

- Each microsporangium is a tubular structure with a multilayered wall. The outermost layer of sporongenous tissue differentiates into a tapetum.
- The sporogenous cells of the microsporangium undergo reduction division and form many haploid microspores.
- Development of microsporangium is of eusporangiate type, *i.e.*, single archesporial cell divides by a periclinal wall forming primary wall cell and primary sporogenous cell.

- The former develops into wall of microsporangium while the latter develops into sporogenous tissue.
- Microsporangium dehisces by means of a longitudinal slit through which microspores are released to start the gametophytic phase.

Female Strobilus: Ovulate Strobilus

- The strobili develop in groups in the axil of leaves of the dwarf shoots **(Fig. 3.7A; Plate II.4).**
- These are much reduced structures, each having a long stalk or peduncle, which bifurcates apically.
- Each bifurcation usually bears a single ovule **(Fig. 3.7B).**
- Out of the two sessile ovules one generally aborts earlier.
- Rarely, three, four or more ovules on a peduncle are also observed. Sprecher (1907) reported an axis with seven ovules, each borne singly on its independent peduncle, and not in pairs.
- At the base of each ovule is present a fleshy cup-like structure called ring or collar.
- The leaves surrounding the ovules do not show bilobed character.

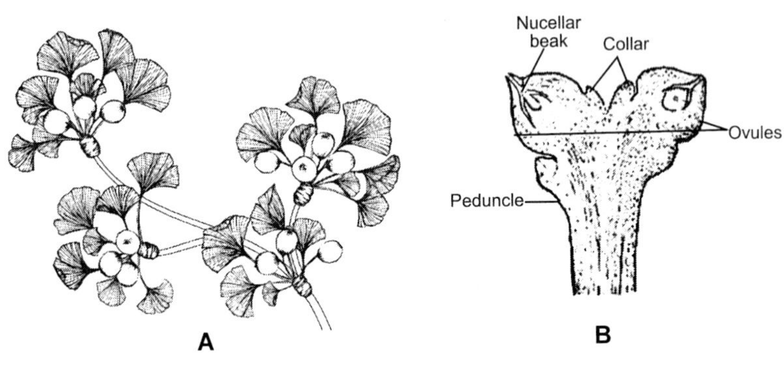

Figure 3.7 A&B: *Ginkgo.* Female reproductive structures.

A - Female strobilus borne on a spur shoot. B - Long peduncle bifurcates and bears an ovule on each of the bifurcation.

Megasporangium: Ovule

- The development of ovule, megasporogenesis and structure of the mature ovule is similar to that of *Cycas*.
- There is one integument in the young ovule but as it matures, a thick integument consisting of three layers, *i.e.*, outer fleshy, middle stony and inner fleshy becomes noticeable.
- A major difference between *Ginkgo* and *Cycas* is that the integuments of *Ginkgo* do not have any noticeable vascular supply.
- Each ovule has a large and prominent nucellus **(Plate II.5).**
- Free apex of the nucellus breaks down into a pollen chamber forming the nucellar beak. The vascular supply of the ovule is ill-developed.
- Two vascular strands enter the inner fleshy layer and reach up to the free part of the nucellus without branching.
- Outer fleshy layer lacks any prominent vascular supply.

Gametophytes

Male Gametophyte: Pollen Grain

- Each microspore is a rounded structure having thin intine and thick exine layers.
- At the time of release, it is a four celled structure with two prothallial cells, a generative cell and a tube cell **(Fig. 3.8A-E).**

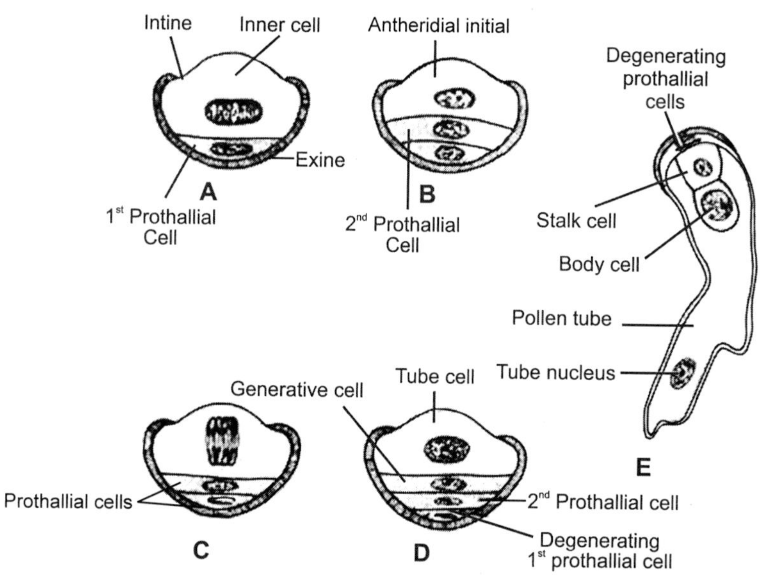

Figure 3.8 A-E: *Ginkgo.* Stages in development of male gametophyte. It is shed at four celled stage.

Female Gametophyte

- Quite deep in the nucellus tissue, a functional megaspore mother cell becomes prominent **(Fig. 3.9A).** The megaspore mother cell develops through meiosis into a tetrad of haploid megaspores, of which only the innermost megaspore remains functional.
- Through mitotic divisions it develops into female gametophyte **(Plate II.5).** Thus begins gametophytic phase.
- The development of female gametophyte in *Ginkgo biloba* has been studied by Carothers (1907). The nucleus of the functional megaspore migrates towards the micropylar end and divides into two followed by a number of free-nuclear divisions forming hundreds of free nuclei.
- A large central vacuole is present in the megaspore at this stage. The wall formation starts gradually and progresses from the periphery towards the centre. During wall formation first the anticlinal walls are formed followed by the vertical walls.
- Each cell generally contains one nucleus but in some cells two or three nuclei are also seen. Due to more rapid cell divisions on the micropylar end a pole-like structure develops. This is called tent pole. The female gametophyte possesses abundant chlorophyll.
- Ultrastructural study of the female gametophyte of *Ginkgo biloba* has been made by Dexheimer (1973). He observed four zones based on food reserves:
    1. Lipid zone,
    2. Starch proteolipid zone with large vacuolated cells,
    3. Starch zone made of large vacuolated cells with peripheral cytoplasm.
    4. Deep or central zone with very little reserve contents.

    Usually two archegonia are formed but rarely three may also be seen **(Fig. 3.9B&C).**

Among extant gymnospermous nectaries are not a common phenomenon, currently being entirely restricted to the gnetophytes, sugar secretion is widespread, particularly in the form of pollination drops (Gelbert and von Adrekas, 2002). Pollination drops appeared very early in their phylogeny. After pollen lands, the pollination drop is withdrawn, thus transporting pollen on the surface of the nucellus (Tomlinson et. al., 1997) where it germinates. In gymnosperms pollination drops typically contain dilute to concentrated carbohydrates with ancillary amino acid components. Although the presented values for gymnosperms exhibit highly variable concentrations of sugars and amino acids, they encompass the range found in angiosperms.

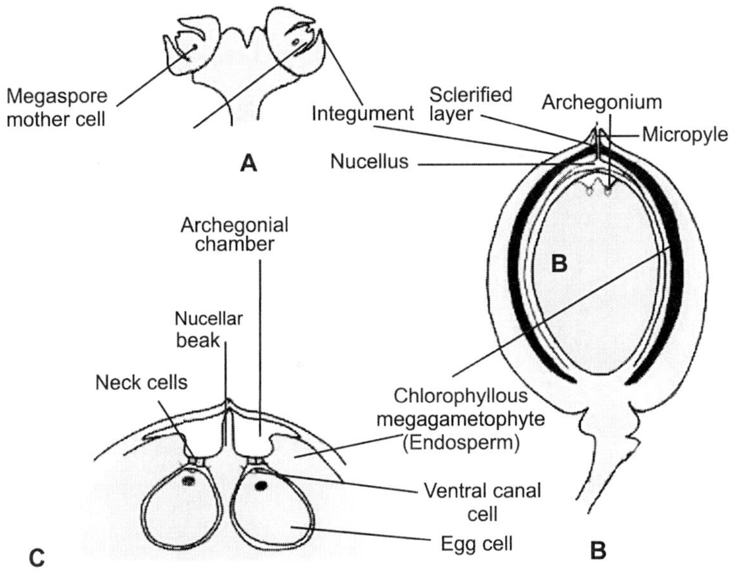

Figure 3.9 A-C: *Ginkgo*. Female reproductive structures.

A - Differentiation of megaspore mother cell. B - The functional megaspore forms a female gametophyte in which archegonia are initiated at the micropylar end. C - Formation of archegonia and nucellar beak.

Archegonium

The development of archegonium starts from the cells towards the micropylar end of the female gametophyte **(Fig. 3.9B&C).** Generally the gametophytic tissue near the archegonia gets absorbed to form an archegonial chamber. According to Favre-Duchartre (1958) the archegonial initials start appearing sometime in June. Each archegonium possesses a short neck made up of only four cells and a small venter having a central cell.

Pollination

- At the time of pollination, the tip of the micropyle secretes a small transparent pollination drop **(Plate II.6)** which is probably formed by the breakdown of cells within the tip of the integument and nucellus to give a mucilaginous liquid.
- Pollen grains that are dispersed by wind, adhere to the pollination drop secreted by the ovule and are sucked in the ovule, where they land in the pollen chamber for further development **(Fig. 3.10A-C).**
- The generative cell divides into a stalk cell and a body cell or spermatogenous cell, where latter divides into two motile spermatozoids. One of the prothallial cells degenerates **(Fig. 3.8A-F).**

- Motile spermatozoids in gymnosperms – *Cycas* and *Ginkgo* were observed for the first time on September 9, 1896 by Hirase, and this discovery is considered as one of the most remarkable events in plant morphology.

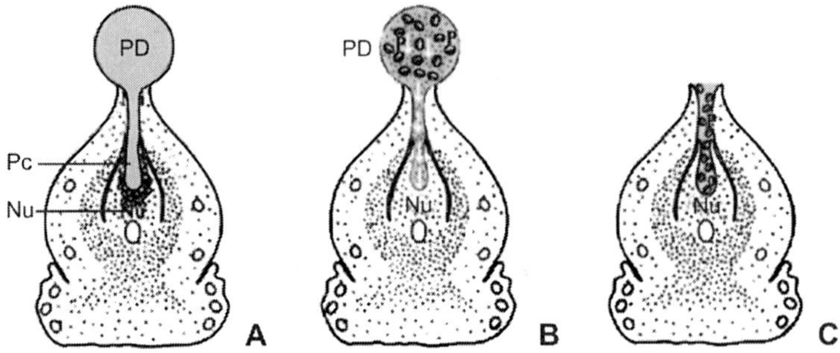

Figure 3.10 A-C: *Ginkgo* Pollination Drop.

A - Ovules with a pollination drop (pd) at the tip of nucellus (nu). B - Pollination drop traps pollen grains. C - The drop along with pollen is taken in the ovule.

Fertilization

- Prior to fertilization from the central cell are formed — a smaller ventral canal cell and a larger egg cell.
- The former degenerates soon to form a cavity variously called 'la chamber receptive'. The cavity is considered to be filled with a liquid in which spermatozoids can swim before bringing about fertilization.
- It is here that the spermatozoids shed off their ciliated cytoplasmic structure.
- Micropyle closes **(Fig. 3.11; Plate II.7)** and fertilization ensues.
- When the spermatozoid is released from the pollen tube, the egg cell forms a small projection which extends between the neck cells.
- As soon as the spermatozoid becomes attached to the egg projection, the latter retreats, leaving a path through the neck.

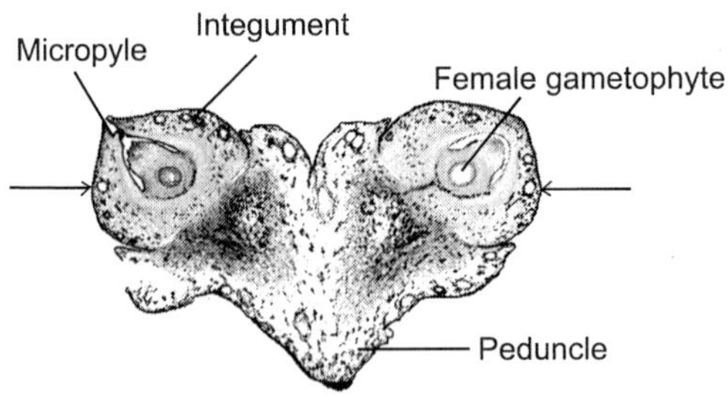

Figure 3.11: *Ginkgo* young ovule shows a distinct micropyle. The integument is undifferentiated into various layers. Note the presence of mucilage cavities in the integument (arrows).

- Only the head of the spermatozoid which includes chromatin, penetrates the egg cytoplasm, where it forms a globular mass. Simultaneously chromatin in the egg nucleus forms another group of granules. Both granules fuse to accomplish fertilization.

Embryogeny

- The first division of the zygote is perpendicular to the axis of the ovule.
- Free nuclear divisions continue till 256 nuclei are formed. The cell wall formation starts just after a few divisions of the nucleus and soon a proembryo is differentiated.
- Three types of cells are clearly differentiated in the proembryo.
- The cells towards the chalazal end are smaller in size with dense cytoplasm, followed by another zone of comparatively larger cells which do not possess such a dense cytoplasm. Elongated cells are present in the uppermost zone.
- There is no differentiation of suspensor in the embryo.
- The embryo differentiates into two cotyledons, plumule, hypocotyl, radicle, and indistinct suspensor tissues **(Fig. 3.12A&B).** Subsequently, the two cotyledons grow rapidly.
- Development of the cotyledon starts from the lowermost zone soon after the differentiation of these zones. Roots, stem and cotyledons are soon differentiated. Generally, five leaves are present in the mature seedling. Out of these five leaves; the first two are decussate with cotyledons while the remaining leaves are irregular in arrangement in embryo and seedling.

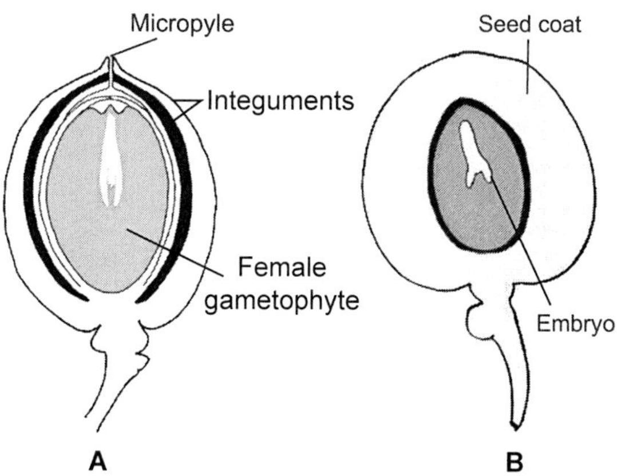

Figure 3.12 A&B: *Ginkgo* Seed.

A - Seed shows a developing embryonal axis. The micropyle is still observed. B - Seed encloses a dicot embryo inside a thick seed coat.

Seed

The ripened seed has a silvery shine (silver apricot nut) fleshy seed coat. When the seed falls on the ground and decays it produces a disagreeable odour as a result of presence of butyric acid, a common by-product of many plants and animals. As many seeds fall on the ground at maturity, cleaning the area of these seeds can be very messy making female trees unfortunately less popular for planting. However, this phase does not last long and in countries like China and Japan where seeds are preferred, regular cleaning of the fallen nuts can solve the problem of bad odour.

Test Your Knowledge

Q 1. Why *Ginkgo* is also called a 'living fossil'?

Q 2. Discuss the morphology of *Ginkgo.*

Q 3. For plantations female trees are not preferred. Why?

Q 4. What has enabled *Ginkgo* plant to survive all harsh conditions over several decades?

Chapter 4

Subclass Gnetidae

The gnetophytes have been treated as separate from other seed plants and have been considered at the ranks of order (Gnetales; Luerss, 1879); class (Gnetopsida; Thom, 1886); division (Gnetophyta; Bessey, 1907), subdivision (Gnetophytina; Cronquist, Takhtajan and Zimmermann ex Reveal, 1996) and subclass (Gnetidae; Cronquist, Takhtajan and Zimmermann ex Reveal, 1996). Among these, the rank of subclass (Christensen *et al.,* 2010) is perhaps most recent in view of the extreme differences between the gnetophytes and the other seed plants, with each of the three subgroups treated as orders; *viz.*

- Welwitschiales Skottsberg ex Reveal, 1993
- Gnetales Blume ex von Martius, 1835
- Ephedrales Dumortier, 1829

The group is a small one, consisting of three families, each with one genus, totalling about 95 species in this treatment. The three gnetophyte genera have largely disjunct distributions. *Ephedra* occurs mostly in semi-arid and arid parts of North America, Mexico, South America (Ecuador to Patagonia and lowland Argentina), South Europe, Asia, and North Africa (including the Canary Islands) (Stevenson, 1993). The other genera are exclusively tropical, *Welwitschia* in Africa, Angola and Namibia, and *Gnetum* in wet forests of Indo-Malaysia, tropical parts of West Africa, Fiji and the northern regions of South America (Schultes and Raffauf, 1990).

Gymnosperms with obscure affinities

For long, gnetophytes were thought to represent a sort of primitive angiosperm. Apparent morphological similarities with angiosperms such as vessel-like water conducting cells, double fertilization and leaf morphology with reticulate venation have historically led to the preposition that gnetophytes form a group that is sister to angiosperms (Anthophyte hypothesis). Superficially they resemble angiosperms more than gymnosperms, and some authors even suggested that they represented a derived group of angiosperms. A molecular study of *Gnetum gnemon* (Winter *et al.*, 1999), however, indicated conifers and gnetophytes as sister groups, and suggested that the evolution of flower-like reproductive structures may have occurred independently in gnetophytes and angiosperms. However, extant species of some gnetophytes differ from conifers in being restricted to warm, mesic habitats while conifers are adapted for cold and water stressed habitats. Though gnetophytes resemble conifers in several aspects of reproduction but resin ducts that occur so abundantly in conifers are absent from gnetophytes. This view has been supported by work of Winters *et al.* (1999) who have produced data of MADS-type floral homeotic genes from *G. gnemon* providing strong molecular evidence for a sister-group relation between gnetophytes and conifers. An independent study of chloroplast DNA from several angiosperm, gymnosperm and gnetophyte taxa (Samigullin *et al.*, 1999) resulted in a similar conclusion, and subsequent studies confirmed this interpretation while also proposing the *Gnetum* and *Welwitschia* clade to be sister to *Ephedra* (Stevens, 2010). The gnetophyte group has invited discussions between evolutionary biologists and molecular biologists, concerning the precise relationship also between the gnetophytes and the other principal groups of gymnosperms. Current thinking, summarized by Stevens (2010), indicates that the cycads first diverged from the protogymnosperm line, followed thereafter by *Ginkgo*, conifers and finally gnetophytes. However, this thinking based almost entirely upon molecular lines of evidence, is open to discussion again in response to studies in future.

The subclass is a small group of gymnospermous vascular plants represented by three living genera: *Ephedra*, *Gnetum*, and *Welwitschia* that exhibit great diversity in form and size among the various species.

Gnetidae living today number ca 65 species in the genus *Ephedra*, 30 or more in *Gnetum*, but only one in *Welwitschia*. The members are distinct from other gymnosperms in possessing long water conducting tubes - vessels, rather than meshing tracheids, an angiospermic character. Most gnetophytes produce naked (exposed) seeds like the other gymnosperms with exception of *Gnetum leyboldii*, which produces seeds enclosed in a juicy, fruit-mimicking layer. Motile spermatozoids are absent from gnetophytes which are heterosporous. Most gnetophytes are dioecious, bearing male and female cones on different plants.

The gnetophyte seed resulting from the fertilization of egg stores its food reserves in a pair of cotyledons, which are the embryonic leaves. *Ephedra* and *Gnetum gnemon* undergo double fertilization although it is unresolved whether double fertilization also occurs in other species of *Gnetum* and in *Welwitschia*. It results in two zygotes. In comparison, double fertilization in many flowering plants results in the production of one zygote and polyploid endosperm. Researchers suggest that the gnetophyte pattern of rudimentary double fertilization with formation of two embryos evolved in a common ancestor of gnetophytes and anthophytes (angiosperms). After divergence of gnetophytes from their common ancestry, the flowering plants may have modified one of the two zygotes resulting from double fertilization to form triploid endosperm, unique to flowering plants. Female gametophyte supplies nourishment to the developing embryo in both *Ephedra* and *Gnetum*. Thus, according to Steeves (1983)- "in development of the angiosperm seed there has been a tendency to shift the dependency of the early embryo directly to the contemporary sporophyte and away from the intermediate gametophye generation or some modified form of that generation."

Gnetum survives in the angiosperm-dominated tropical rain forest even though individual plants are widely spaced, as are individual plants of angiosperm species. At least some species of *Gnetum*, *Ephedra* and *Welwitschia* produce nectar, which enhances the efficiency of pollination. Insects pollinate *Gnetum*; wind and insects probably pollinate the other gnetophytes. *Gnetum leyboldii* seed is covered by a fleshy integument layer called a pseudofruit. Its sweet flesh entices the chestnut-mandible toucan to feed on the pseudofruit. Later the toucan regurgitates the seed, sometimes dispersing it.

Molecular evidence suggests that gnetophytes are a phylum closest to flowering plants; data from ribosomal RNA are supported by ribulose-1,5-bisphosphate carboxylase (*rbcL*) gene sequence information. Other similarities between gnetophytes and flowering plants (especially dicots) such as vessels in the wood, seeds with two cotyledons, and leaves with net-like veins (in *Gnetum*) have been already mentioned.

Order Gnetales

It is a monotypic order with family Gnetaceae where the members are woody climbers, a few species erect, branches jointed and thickened at nodes. Leaves are simple, opposite and show reticulate venation. Strobili are unisexual or imperfectly bisexual. Each staminate shoot consists of a pair of fused bracteoles, known as perianth. Each ovule possesses three layers of envelopes.

Gnetum

Distribution

The genus includes about 30 species that are distributed in tropical and humid areas of the globe *i.e.*, Africa, Asia and South America. No species are reported from Australia. Five species, *G. gnemon*, *G. latifolium*, *G. montanum*, *G. contractum*, and *G. ula* are reported from India.

In India, the plant is distributed in the Himalayan region. Out of the many species, *G. gnemon* is a small tree or shrub with short green drooping branches while *G. latifolium* is either an evergreen shrub or a climber. *Gnetum ula* and *G. montanum* are hardy climbers.

Sporophyte

Morphology

- The shrub or small tree of *Gnetum* shows shoots which are dimorphic but the difference between the two *i.e.*, long and short shoots is not well demarcated.
- The short shoots borne on long shoots bear about eight pairs of opposite and decussate foliage leaves while the long shoots may also show similar leaves or the leaves may be highly reduced, represented by scales only.
- Foliage leaves are simple, with short petiole, and opposite and decussate arrangement. The venation is reticulate as in angiosperms **(Fig. 4.1).**
- Root is tap root which gives way to lateral roots.

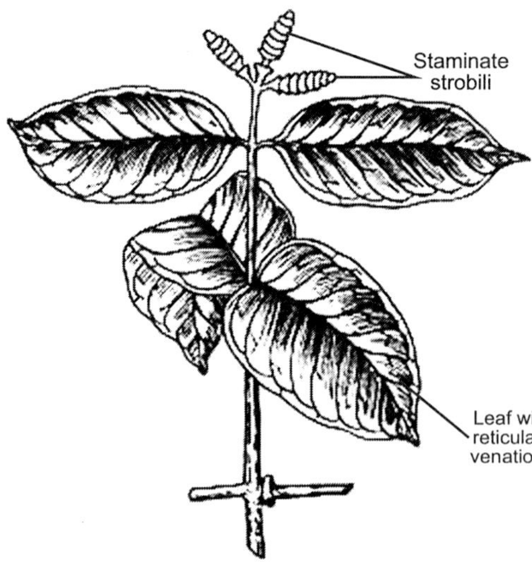

Figure 4.1: *Gnetum* Sporophyte.

A reproductive axis showing male inflorescence. The sub-sessile opposite and decussate leaves bear resemblance to angiosperms.

Anatomy

Stem

- Resembles that of an angiosperm in construction and has epidermis with cuticle and sunken stomata.
- Cortex is large parenchymatous with sclereids and spicular cells (cells with calcium oxalate crystals) in the tissue.
- Vascular bundles, about twenty are enclosed in indistinct pericycle and endodermis. They are conjoint, collateral, open, endarch and are arranged in a ring **(Fig. 4.2A).**
- The bundles are separated by distinct medullary rays; xylem has vessels **(Fig. 4.2B)** and tracheids, latter are with uniseriate bordered pits **(Plate III.1A).**

Secondary Growth

In some species as *Gnetum gnemon* normal secondary growth occurs while in *G. ula* and *G. africanum* secondary growth occurs by several successive rings of cambia which develop one after other **(Fig. 4.2B)**. One ring of cambium arises and cuts of secondary xylem and secondary phloem. But soon its activity ceases and another ring of cambium arises with similar activity. Thus, the cortical region shows several rings with small secondary tissue. In several other species, the entire ring may not remain active *i.e.*, the ring of cambium is formed but it cuts off secondary tissues

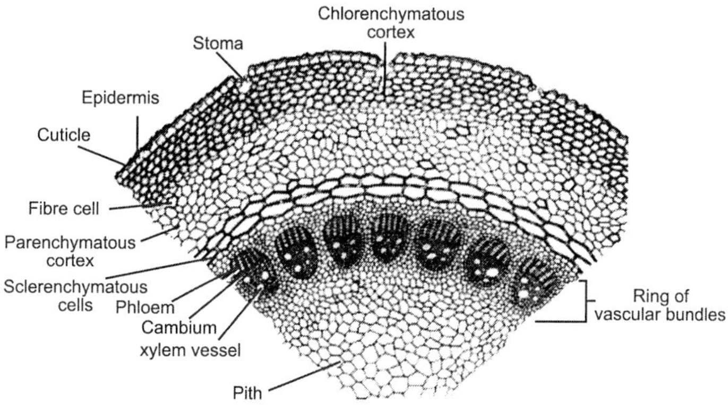

A

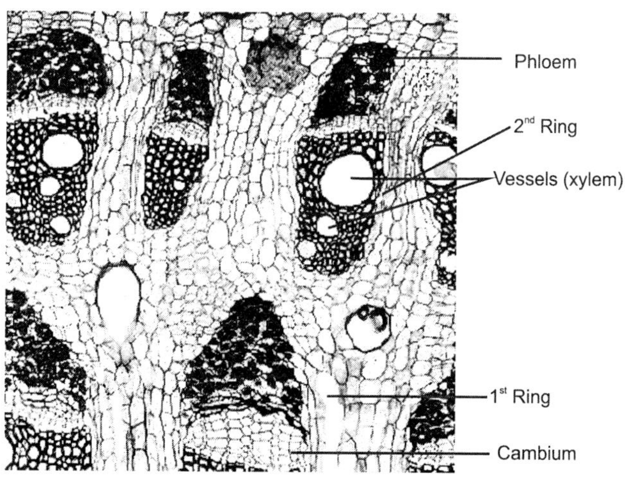

B

Figure 4.2 A&B: *Gnetum* T.S. Old Stem.

A - Portion to show general construction with a ring of vascular bundles. B - Presence of vessels in wood. Also seen are rings (at least two) of secondary tissues along with short-lived inter fascicular cambium.

on one side only, resulting in an eccentric secondary ring **(Fig. 4.3A&B)**. Similar to this anomaly, the cork cambium also shows irregular growth and the secondary tissue formation is anomalous as a result of this activity. The secondary xylem vessels show a single large perforation. A ring of sclerotic cells is also seen in the cortical area **(Plate III.1B).**

The secondary growth resembles that of *Cycas* in abnormalities of vascular and cork cambial activities.

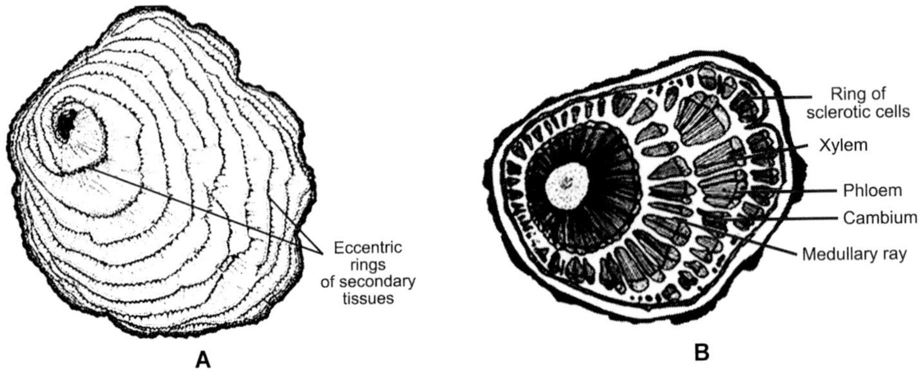

Figure 4.3 A&B: *Gnetum* T.S. Old Stem.

A - Outline diagram to show abnormal secondary growth with eccentric secondary rings. B - Diagrammatic sketch to show abnormal secondary growth results in wedge-shaped vascular bundles.

Root

- Epidermis or epiblema is single layered with thin cuticle.
- Cortex multilayered, parenchymatous, cells are polygonal with starch. Fibres lie interspersed.
- Endodermis is single layered and is followed by a pericycle.
- The vascular bundles have xylem and phloem, with tracheids and vessels in xylem. Phloem lacks companion cells.
- Pith occupies the centre.
- Secondary growth results in formation of secondary xylem, secondary phloem and cork cambium.

Leaf

- The outline is undulated **(Fig. 4.4A)** and outermost is epidermis with cuticle; the lower epidermis has stomata which are of the sunken type.
- Mesophyll is differentiated into palisade and spongy tissue, thus, making leaf dorsiventral in construction. Interspersed in the latter are sclerotic cells **(Plate III.2A)**.
- Vascular bundles are conjoint, collateral and endarch, with xylem facing upper epidermis.
- A ring of sclerenchymatous tissue is associated with vascular bundles **(Fig. 4.4B; Plate III.2B)**.

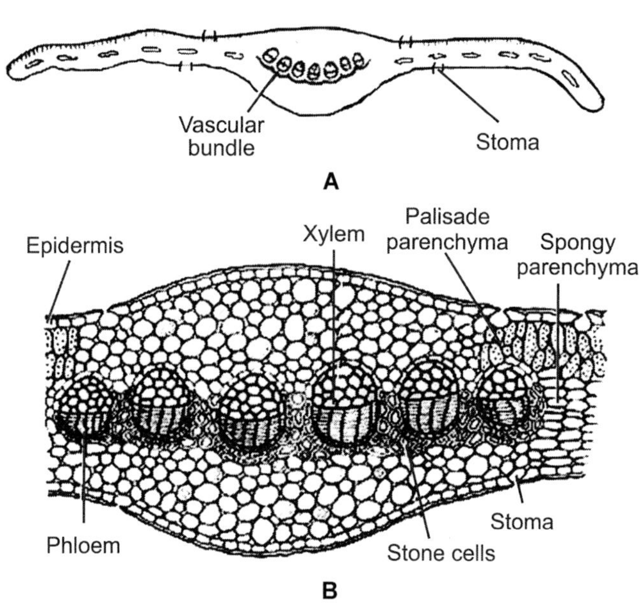

Figure 4.4 A&B: ***Gnetum*** **V.S. Leaf.**

A - Outline diagram to show thin wings and a broad midrib. B - Portion to show several vascular bundles with associated stone cells. Stomata occur on both surfaces.

Reproduction

Vegetative Propagation

Since the climbers are eaten by the local people, vegetative propagation of plant is common. In gardens it can easily be multiplied by stem cuttings.

Sexual Reproduction

The reproductive structures bear resemblance to angiospermic structures and are frequently referred to as 'flowers'.

Strobilus (Inflorescence)

- Plants are dioecious, however, in rare cases they may be bisexual.
- Strobilus arises in axil of a leaf of short shoot; may terminate the growth of the shoot.
- Male strobilus is usually branched while female is unbranched.
- In general a strobilus has an axis with two sterile connate and opposite bracts at the base. Above these is a series of circular bracts superposed one above other referred to as cupule or collar. Occasionally upper one or two collars may be reduced and sterile.

- In the young strobilus two basal bracts enclose the entire axis and as the internodes do not elongate, the collars appear contiguous.

Male Strobilus

- Each strobilus shows a series of collars **(Plate III.3A&B).** Above each collar there are four to six rings of twelve to fifteen male flowers with basipetal development **(Fig. 4.5A-C).** A ring of seven to twelve abortive ovules is seen in the collars that are towards the apex. The male cone contains sterile ovules (female parts), suggesting that functional male and female reproductive structures once resided on the same plant.
- The entire structure is connected to twelve to fourteen bundles in the peduncle of the male branch.

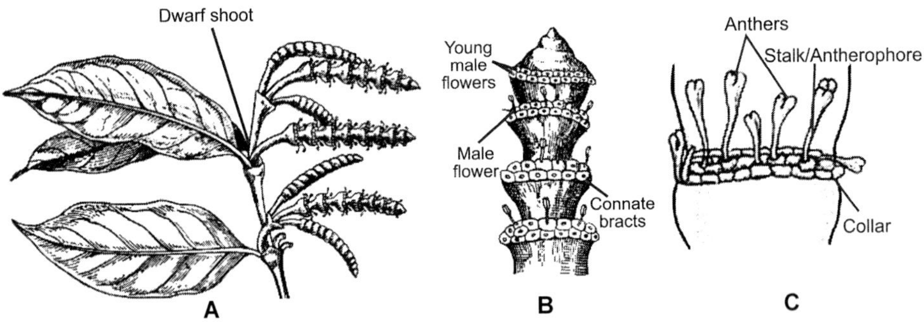

Figure 4.5. A-C: *Gnetum*. Male reproductive structures.

A - Portion of a twig from male plant showing clusters of branched inflorescence. B - Single strobilus with collars bearing male flowers. C - Single collar enlarged to show male flowers arranged in rows on the collar.

Male Flower

- Consists of a stalk, the antherophore that bears two anthers enclosed by a perianth.
- At maturity, the antherophore elongates and anthers emerge out of the perianth **(Fig. 4.6).**
- Variations are seen in the flowers – adjacent flowers may fuse, flowers of different rings may fuse bearing more than one antherophore inside the perianth, the number of anthers per antherophore may also be more than two.
- Variations are also seen in strobilus. The ring of sterile ovules may start to develop and produce abortive structures.
- In certain specimens, the lowest collar may be sterile or it may be partially fertile with male flowers on one side only.

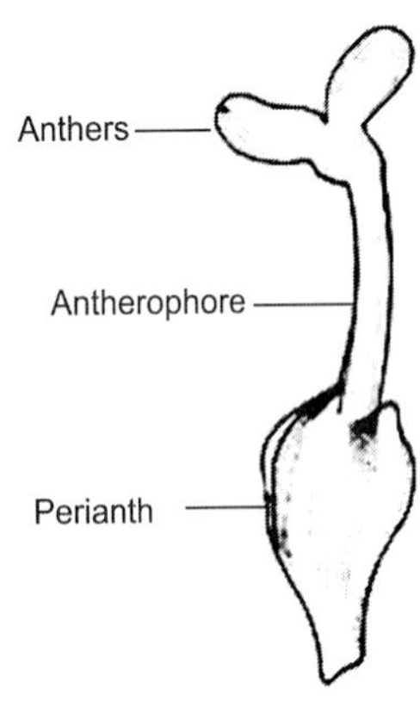

Figure 4.6: *Gnetum* male flower shows a distinct antherophore, perianth and anthers.

- The number of collars may be less, there may be elimination of male flowers and the female flowers may then become fertile.
- The number of rings bearing male flowers or the number of male flowers per ring may be less. In these, ovules tend to develop normally.
- There may be two rings bearing ovules rather than one.
- Secondary male strobili may develop from lower collars. However, these strobili remain stunted and bear only few collars.
- The axis may not be straight but is curved, rarely two strobili may fuse along part of their length.

Microsporangium

- Each male flower has two unilocular microsporangia on a stalk. The entire structure is enclosed in a sheath or a perianth. Once the microspores are ready for release, the axis elongates and the microsporangia are pushed out of the perianth **(Plate III.4A&B, 5A&B).**
- Has epidermis, middle layers and tapetum, however, it is only the epidermis which persists at maturity while middle layers are first to disintegrate. Tapetal cells become two-four nucleate and may also show polyploid condition.

Female Strobilus

- Has a ring of four to seven ovules above each collar.
- Initially all ovules are at same stage of development but soon some cease to develop and thus, abort **(Fig. 4.7A-C).** A few then form seeds and the axis becomes thick in order to withstand the weight of the seeds **(Plate III.6).**
- The female strobilus is also fed with twelve to fourteen bundles.

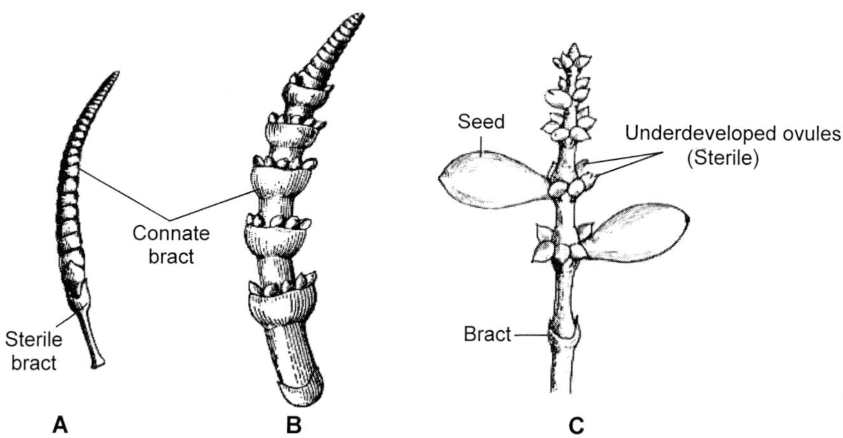

Figure 4.7 A-C: *Gnetum*. Female reproductive structures.

A - Female strobilus when young has closely placed collars. B - Single collar with a few developing ovules. C - Only one ovule reaches maturity .

Ovule

- The ovules are arranged around the axis **(Plate III.7).**
- Ovule comprises a nucellus surrounded by three envelopes. In abnormal cases there may be a fourth underdeveloped integument or there may be just two integuments.
- The outer envelope or integument covers the ovule completely and is massive with numerous sclereids and laticiferous ducts. Stomata are also seen. It is more commonly known as the perianth **(Plate III. 8A).**
- Middle envelope or integument is not massive but in older ovules its upper portion becomes swollen. This integument also has laticiferous ducts, sclereids and stomata. It is generally accepted as the outer integument **(Plate III.8B).**
- The inner envelope or integument arises as an annular primordium and forms several lobes at the stage of pollination. It develops two proliferating structures from its middle portion, *viz.*, a flange and a micropyle-closing tissue **(Fig. 4.8).** It remains fused with the nucellus at the base and is thin extending into a micropyle, which disintegrates after pollination. It is more commonly known as the inner integument.
- Formation of the flange and the micropyle-closing tissue is an interesting feature. At about the stage of pollination the inner envelope develops an umbrella-shaped structure called a flange. The initiation of the flange starts as a slight swelling at the level just above the tapering ring of the middle envelope. It grows downwards to cover the apex of the middle envelope and also upwards to form a small process circular in outline. The flange usually has cellular protrusions from its outer layer developing into multicellular hairs. These hairs seek their way in an irregular fashion between the outer and middle envelopes. The entire outer surface of the flange becomes tightly appressed against the outer envelope in the subsequent stages. Formation of the flange begins by periclinal divisions in cells of the outer dermal layer, their derivatives usually first dividing in the same plane, but more irregularly later on. The upwardly directed flange is only one to three cell layers thick. A similar, but slighter, proliferation of the outer dermal layer of the inner envelope also takes place at the level where this envelope is in close contact with the middle one. At the level of the flange the micropylar canal becomes closed **(Plate III. 8A&B)** (Takaso and Bouman, 1986).
- Twin ovules have been also observed as an abnormality but only one matures.

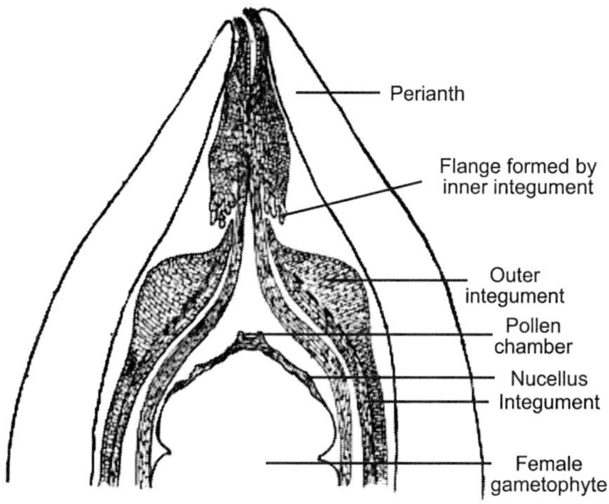

Figure 4.8: ***Gnetum*** **L.S. Ovule to show three integuments and a pollen chamber. The perianth covers the entire structure, outer integument is swollen towards micropyle and inner integument forms the flange towards the micropyle.**

Megasporogenesis

Initially one or two archesporial cells differentiate in the hypodermal region and divide to form an outer primary parietal cell and inner sporogenous cell. These cells may further divide mitotically as a result of which older ovules show twelve to twenty sporogenous cells arranged in a row. The sporogenous cells then act as megaspore mother cells. Each mother cell undergoes a meiotic division not accompanied by cytokinesis and a tetrasporic (coenomegaspore) development occurs. Not all the mother cells divide, some may degenerate early in development.

Gametophytes

Male Gametophyte: Pollen Grain

- Has a thin intine and thick exine and at maturity has three nuclei – a larger comparatively poor in chromatin is the tube nucleus, a small with densely stained nucleus and a cytoplasmic sheath is the generative cell and a still smaller and darkly stained one is prothallial cell. Pollen is shed at this three celled stage.
- Four nucleate, and double pollen grains were abnormalities encountered.
- The prothallial cell never enters the tube and simply degenerates while tube nucleus is the first to enter.
- The generative cell divides later to produce two sperm nuclei.

Female Gametophyte

- A female gametophyte starts to develop and nucellar cells below the four nucleate embryo sac show dense cytoplasm and get arranged in radiating rows. This fan-shaped tissue was termed pavement tissue by Coulter 1908. It is perhaps a source of nutrients to the gametophyte.

- About six to ten coenomegaspores start growing into female gametophyte but only one near the chalazal end reaches maturity.
- Rarely if the surviving gametophytes continue to develop, they may reach only up to 32 nucleate stage.
- The gametophyte which is initially ovoid in outline, now becomes broader at the micropylar end. Further development results in embryo sac containing 128, 256 and 512 nuclei **(Fig. 4.9A-D).**

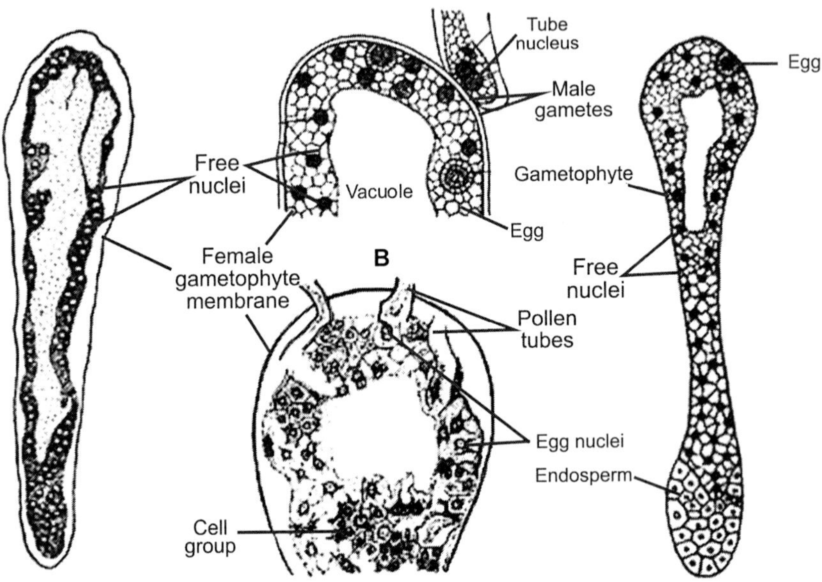

Figure 4.9 A-D: *Gnetum.* Stages in development of female gametophyte.

A - Free nuclear stage. The nuclei are arranged towards periphery and the micropylar end is broader than the chalazal end. B - Shows potential egg nuclei near the tip of the pollen tube. C - Several egg nuclei are seen in the gametophyte. D - Chalazal part is seen to cellularize while eggs still differentiate.

Pollination

- In several ways, dioecious *Gnetum* plants resemble many angiosperms especially in insect pollination and seed dispersal by fruit-eating birds. At least some species of *Gnetum* (and also *Ephedra*, and *Welwitschia*) produce nectar, which enhances the efficiency of insect pollination.
- The female gametophyte with about 32-64 nuclei is considered to be mature to receive the pollen. The lacerated micropyle of the gametophyte receives pollen which start to germinate here.

Egg

- There are no archegonia and the eggs differentiate in the embryo sac once the mitotic divisions are over.
- Simultaneously, as the pollen tube starts growing inside the embryo sac, some of the nuclei which come in contact with the pollen tube start to grow rapidly, increasing in size. These nuclei get a layer of cytoplasm around them and act as eggs which upon fertilization give rise to the zygotes.

Fertilization

Double fertilization has been reported only in *Gnetum gnemon* and no other species of *Gnetum*. *Gnetum gnemon* (and perhaps *Welwitschia*) does not form egg cells; instead, as a product of meiosis, it forms egg nuclei that are free in the female gametophyte. Each ovule forms a pollination droplet that retracts, pulling pollen into the ovule. In *Gnetum*, each of two sperm nuclei from one pollen tube fertilizes a separate female (haploid) nucleus or the egg nucleus; two separate fertilizations form two (diploid) embryos. Ultimately, just one of the two embryos reaches maturity in each seed. Subsequent to fertilization, several female nuclei fuse, developing polyploid, embryo-nourishing tissue that is the functional equivalent of endosperm but of different origin.

In a study on *G. gnemon*, within seven days of pollination, as many as six pollen tubes, each carrying two sperm nuclei (derived from a division of generative cell), enter the female gametophyte and protrude into the central vacuole, of coenocytic female cytoplasm. Soon the free nuclei migrate around the apex of each pollen tube. By this time, chromatin within each sperm nucleus is organized in a characteristic crescent-shaped pattern. Two sperm nuclei from each pollen tube are then discharged into the surrounding free nuclear female cytoplasm each fertilizing a female nucleus to form a zygote. In case of multiple pollen tubes entering a mature female gametophyte, most, if not all, participate in double fertilization events. The tube (vegetative) nucleus remains within pollen tube and does not participate in fertilization process. Prior to fertilization none of the free female nuclei that migrate around a pollen tube differentiates as an egg cell, and all appear to represent potential female gametes. Before fertilization, these female nuclei are equivalent in DNA content and position within the cell cycle. By the time sperm nuclei are discharged, free nuclear divisions have permanently ceased within the female gametophyte, and female nuclei located near a pollen tube contain the 2C quantity of DNA. Because these female gametophyte nuclei are indistinguishable from a structural and cell cycle perspective, it is likely that all are equally fertile and that fertilization of these may merely be by chance or proximity to the region at which sperm nuclei are discharged from a pollen tube (Carmichael and Friedman, 1995). Wall formation ensues first in the chalazal part of the gametophyte, proceeding towards the neck. The cytoplasm compartmentalizes with 4-10 nuclei per compartment. Later, the nuclei in the upper portion may degenerate and thus, the lower portion of the gametophyte contributes to endosperm. Nuclear fusion in each compartment results in a polyploid condition. The nuclear fusion proceeds generally from base upwards. The upper portion has more cells dividing and the gametophyte becomes much broader here. The endosperm slowly fills the entire gametophyte with proembryonal tubes in the upper portion. The centre of activity in endosperm is central axial part. Rarely the endosperm starts to cellularize very early in development when even the eggs have not differentiated.

Embryogeny

Germination of zygote and embryo formation.

- Generally two to four zygotes are seen in a gametophyte. The zygote either without dividing further or after one division may give out a tube variously termed proembryonal tubes, embryonal tubes, suspensor tubes and primary suspensor tubes.
- Generally a single tube is produced from a zygote and the nucleus then moves into it. The extra tubes without the nuclei degenerate.
- These tubes grow downwards while the endosperm grows upwards. These tubes after travelling two thirds of the endosperm length start aggregating in the central core while the surrounding cells breakdown forming a lacuna.
- Embryo development is initiated at the tips of only a few embryonal tubes.
- The tips show embryonal mass, secondary suspensor cells and primary suspensor cells **(Fig 4.10A-D).**
- The primary suspensor tubes cease to elongate but the secondary suspensor cells divide and form elongate folded tubes that push the embryonal mass into the nutritive tissue **(Fig. 4.10A&B).**

Polyembryony

- The central core of the endosperm shows a bunch of secondary suspensors and embryonal masses formed as a result of polyembryony. Though a number of embryos start to develop only one reaches maturity.
- Mass of cells at the tip of a primary suspensor becomes bilobed and each lobe gives rise to an embryo.
- The primary suspensor may produce a chain of cells which grow laterally into embryo.
- A cleavage of embryonal mass may occur due to disintegration of some of its cells and each small group gives rise to an embryo.

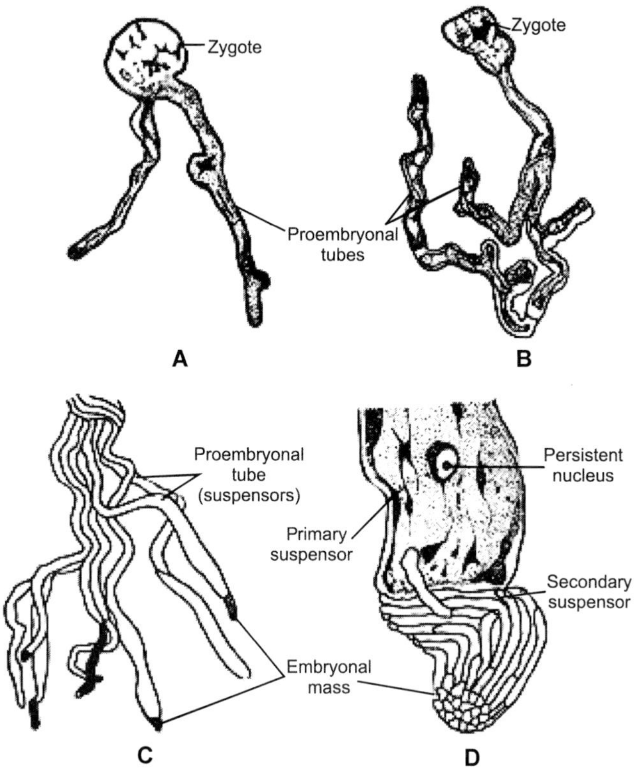

Figure 4.10 A-D: *Gnetum*. Stages in embryogeny.

A&B - Proembryonal tubes differentiate in early stages. C - The suspensor cells differentiate which elongate pushing embryonal cells into nutritive tissue. D - The embryonal mass differentiates at the tips of these tubes.

Seed

- Seed is oval in outline and three envelopes remain free except at the upper side of the flange where it is completely fused with the outer envelope and at the lower side where it is partly fused with the apical part of the middle envelope. The nucellus persists as a small projection. The endosperm fills the greater part of the seed.
- The three envelopes differentiate into the fleshy outer, sclerenchymatous middle and compressed inner layers of the seed coat.
- The embryonal mass gives rise to two cotyledons enclosing a shoot tip, the root tip differentiating on the opposite direction.
- After the stem and root tips are well developed, a small hump arises on one side of the hypocotyl. This is known as feeder.

- Rest of the growth is completed after seeds fall to the ground. The life cycle has distinct stages in sporophytic and gametophytic generations.

Life cycle of *Gnetum* is depicted in Figure 4.11.

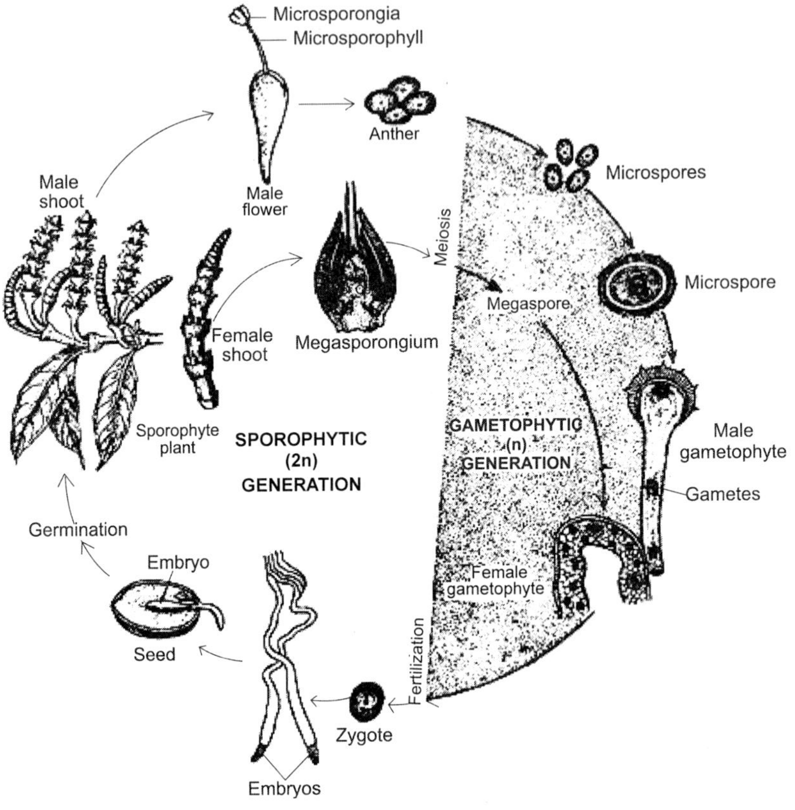

Figure 4.11: *Gnetum* life cycle.

Relationships of Gnetales

The members show important differences from angiosperms:

- Xylem tracheids have bordered pits and phloem lacks companion cells.
- The ovules are naked and are borne in strobili.
- Pollination is an ovular function and no true style or stigma is present.
- Strobili are typical of gymnosperms.
- Microspore germination is gymnospermous.
- No fruit formation.
- Polyembryony is common.

The main characters on the basis of which the three genera (*Ephedra, Welwitschia* and *Gnetum*) were grouped together are:

- Both tracheids and vessels are present.
- The ovules and microsporangia are borne on fertile shoots resembling flowers, which are organized into compound strobili.
- The ovules are enclosed within coverings additional to the so-called integument.
- The upper part of the inner envelope is prolonged into a long tubular structure called the micropylar tube.

Inspite of all these similarities, Florin (1931, 33) and Eames (1952) have given strong support for a breakup of the order Gnetales and the establishment of three separate orders, Ephedrales, Welwitschiales and Gnetales. It is worthwhile to know similarities and differences.

Gnetum and *Ephedra* differ in several features **(Fig. 4.12).**

- They differ in habitat: *Ephedra* is small and bushy while *Gnetum* is a tree or a liana.
- *Ephedra* grows in dry habitat while *Gnetum* grows in humid tropical or sub-tropical regions.
- Leaves are small with parallel venation in *Ephedra* and in *Gnetum* they are reticulate.
- The stem in *Ephedra* is photosynthetic with no secondary growth but in *Gnetum* it is with anomalous secondary growth.
- The vessels in *Ephedra* have a number of bordered pits on end walls while in *Gnetum* a large perforation is seen.
- Strobili in *Ephedra* have a strong homology between male and female. *Ephedra* also shows a tent pole in the micropylar region.
- In *Ephedra*, prothallial cells are seen in pollen grain development while these are absent in *Gnetum* species.
- *Ephedra* has a monosporic and *Gnetum* has a tetrasporic embryo sac.
- Pollen chamber in *Ephedra* is deep and wide while in *Gnetum* it is shallow.
- In *Ephedra* the female gametophyte gets cellularized completely before fertilization but in *Gnetum* only the lower portion is cellularized pre-fertilization.
- Archegonia are present in *Ephedra* but absent in *Gnetum.*
- Embryo in *Ephedra* has no feeder but in *Gnetum* it is seen.

For convenience the reduction in the number of cells that constitute the male gametophytes of gymnosperms particularly conifers and gnetophytes is represented through line diagram* *(Fig. 4.12A-E).***

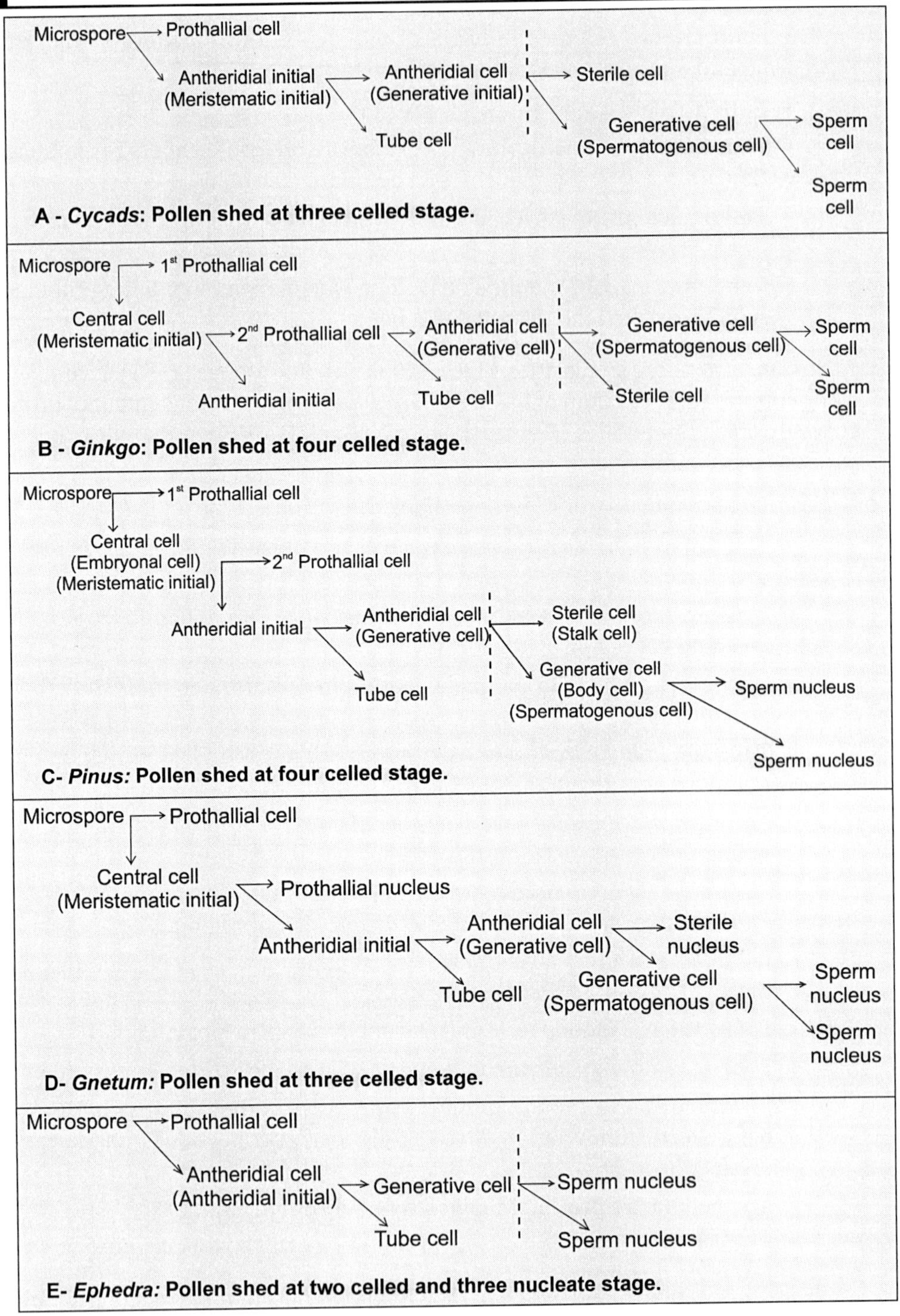

Figure 4.12 A-E: *Gnetum*. Male gametophyte development-Comparison with: A - *Cycads*, B - *Ginkgo*, C - *Pinus*, D - *Gnetum*, E - *Ephedra* (Modified after Fernando *et al.* 2010).

Gnetum and Angiosperms

Gnetum has long held a key position in any discussion relating to the origin of angiosperms. According to Arber and Parkin (1908), the order Gnetales is regarded as a race of gymnosperms, most nearly related to the angiosperms. Both the groups are considered to have sprung from a common stock (the so-called Hemiangiosperms) and in many respects their lines of development continued parallel to each other. Thompson (1916) is of the view that the ancestors of angiosperms were not far removed from the genus *Gnetum* itself. Still others go so far as to say that *Gnetum* actually belongs to the angiosperms and a close relationship has been observed between Piperaceae and Gnetales.

Some features of *Gnetum* show an approach to the angiosperms.

- Habitat and foliage show remarkable resemblances; leaves are broad and netted venation.
- Shoot apex has a angiospermic organization.
- Presence of vessels in xylem.
- The Maule reaction specific for angiospermic wood is also shown though weakly by *Gnetum*. (Maule reaction – chlorination of lignin by treatment with potassium permanganate and strong HCl followed by NH_3 gives a bright red colour).
- Homologies with reproductive organs of *Gnetum*, *Welwitschia*, *Ephedra* and angiosperms occur. Several floral types among angiosperms are derived directly from the condition in *Gnetum* (similar to angiospermic flower as in *Liriodendron*).
- Male and female flowers show perianth.
- The micropylar tube acts as a style in which pollen grains germinate.
- Tetrasporic type of embryo sac formation is seen and the female gametophyte does not have any archegonia.
- At maturity egg cells are seen to differentiate in the female gametophyte. This has resemblance to the embryo sac of *Plumbago* where there are no synergids and one egg cell occupies the micropylar end. While the lower portion is cellular, the upper portion of the female gametophyte remains free nuclear.

Thompson has studied *Gnetum* with respect to the affinities in detail. The pollen grain does not have a prothallial cell as in so many gymnosperms. Therefore, *Gnetum* has completely departed from the gymnosperm condition and has arrived at the angiosperm condition **(Fig. 4.12).** According to him not all pollen grains germinate in the 'style'. The fact that pollen grains germinate at a distance form nucellus closely connects *Gnetum* with angiosperms and removes it from gymnosperms.

**From his assessment of the condition in Gnetum, Thompson (1916) had earlier concluded "almost every structure in Gnetum shows some approach to the angiospermic condition and that some structures show conditions almost completely of highly evolved flowering plants. Such a body of evidence from all sides can scarcely be ignored or put aside as a result of parallel development. Sum of evidence from all sides seems to lead to the conclusion that angiosperms are phylogenetically related to Gnetales and the ancestors of angiosperms were not far removed from the genus Gnetum. The genus however, remains largely a phylogenetic puzzle. It is gymnospermous, but possesses some strong angiospermic features".*

Economic Importance

Gnetum gnemon is cultivated in Malaysia, Indonesia and some south east Asian islands for its edible fruits. The yellowish-red outer part of the fruit is discarded and the seeds are eaten after roasting or cooking; the seed kernel is mashed, moulded into cakes or biscuits, dried in sun and fried in boiling oil. The young leaves and strobili are eaten as vegetable. The bark yields a fibre which is obtained by peeling the branches and beating and splitting the bark into fine filaments. The fibre is strong and durable in seawater with a good tensile strength both in dry and wet condition. It is valued for fishing nets and lines. Ropes made from it are strong, pliable and light. It is also suitable for paper pulp. The kernel oil of *G. ula* is used for relief from rheumatism in Kerala. *Gnetum montanum* is reported to possess piscicidal properties.

Test Your Knowledge

Q 1. Why is *Gnetum* called a synthetic genus?

Q 2. List the similarities between *Gnetum* and angiosperms.

Q 3. What is the similarity between secondary growth of *Cycas* stem and *Gnetum* stem?

Q 4. Describe the male and female flower in *Gnetum*.

Q 5. Define: Flange, pavement tissue, coenomegaspore, collar.

Q 6. Draw L.S. of *Gnetum* Seed.

Ephedrales

A monotypic order includes plants that may be herbs, woody shrubs or lianas. The stem is jointed with distinct nodes and internodes, is photosynthetic with scale-like leaves. Secondary wood has vessels. The plants are dioecious with compound male and female strobili. The ovule is surrounded by two envelopes where the inner projects as a long tube. Archegonia are present as in most of the gymnosperms. The embryo is dicotyledonous.

Ephedraceae

The monotypic family has been around since the early Cretaceous period, with at least one genus (*Ephedrispermum*) in addition to *Ephedra* (Rydin *et al.,* 2006). The surviving species of present times seem to belong to a relict group that has survived climate changes prevailing since Oligocene, having resulted in development of very extensive semi-arid habitats.

Ephedra

Common names for *Ephedra* in English include Joint pine, Joint fir, Mormon tea or Brigham Tea. The Chinese name for the *Ephedra* species is *mahuang* (*máhuáng*; ma-huang; literally meaning "cannabis yellow"), cañatilla, popotillo, tepopote (Stevenson, 1993). *Ephedra* is also sometimes called sea grape (from the French *raisin de mer*), a common name for the flowering plant *Coccoloba uvifera* (see Ephedrales by World Heritage Encyclopedia).

Distribution

Approximately 65 species of *Ephedra* grow in dry climates and are about equally distributed between the Old World and New World. They occur over diverse areas of the Northern Hemisphere, including south western North America, Europe, north Africa, and south west and central Asia, and, in the Southern Hemisphere, in South America south to Patagonia. In temperate climates, *Ephedra* species generally grow on shores or in sandy soils with direct sun. In India, the species are generally shrubs distributed in regions of Kashmir, Sikkim excepting *E. foliata* which is a climber distributed in plains of Punjab and Rajasthan.

In *Ephedra* the habit varies greatly even in the same species. *Ephedra distachya* is a dwarf bush, a few inches high. *Ephedra distachya* and other species are also represented by tall bushes up to 6 ft. in height, with virgate branches. *Ephedra triandra* may attain habit of a small tree with a trunk approximately a foot in diameter. Climbers of the "weaver" type with slender pendent branches are found in *E. altissima* and *E. fragilis*; these and other species include forms in which the branches are prostrate. *Ephedra trifurca* is a prostrate species with underground rhizome.

Most species, especially growing in loose soils, spread by means of branching rhizomes which arise from buried nodes of the erect stems; they thus, exert a binding effect upon the soils in which they grow. In *E. alata*, growing in the sand dunes of the Areg formation of Algeria and Tunis, the rhizomes attain a length of several metres and form efficient sand binders. Normal erect shoots arise from the axils of younger leaf sheaths. When these shoots are isolated by the breaking or decay of the parent rhizome they differ from seedlings only in the absence of tap-roots (Pearson, 1929; see also www.conifers.org/ep/Ephedraceae.php).

Ephedra is the largest of the three extant genera included in the group of gnetophytes. Based on the plesiomorphic or primitive character states of ovule, embryo and gametophyte development, it is considered the most basal genus (Crane, 1984; Doyle and Donoghue, 1986, 1992; Doyle, 1996; Hickey and Taylor, 1996; Price, 1996). The taxonomy of *Ephedra*, an arid adapted genus, has always been controversial due to the highly reduced morphology and the convergent evolution of species occurring in xeric habitats.

Do You Know?

Several incidences of pollen dimorphism have been observed for *Ephedra*. Beug (1956) described both normal grains and grains considerably larger and with a greater number of ridges as occurring together. This size difference was attributed to an incomplete separation of pollen mother cells during meiosis. The larger grains could also represent macropollen, possibly with unreduced gametes and be indicative of polyploidy in the genus.

Sporophyte

The genus is named from the "Greek ep-, upon, and hédra, seat or sitting upon a place; from the ancient name used by Pliny for *Equisetum*; the stems resemble the jointed stems of *Equisetum*, the segments of which appear to sit one upon the other" (Stevenson, 1993). At first glance, *Ephedra* resembles joint-stemmed horsetails, but *Ephedra*'s shrubby habit and strobili distinguish it from horsetails **(Plate IV.1)**.

Morphology

It is characterized by greenish, slender, jointed branches that are longitudinally ribbed. Scale-like leaves that arise from the node and are borne in opposite decussate pairs, sheath the nodes, where two stem joints meet. The reduced leaf surface is probably a response to water loss in semi-arid regions. The plants are

A

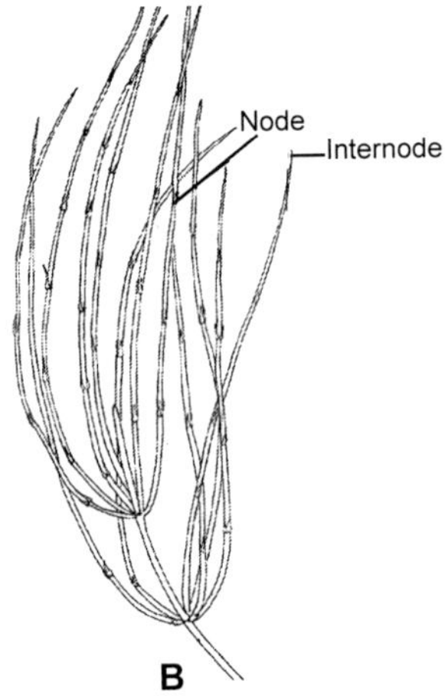

B

Figure 4.13 A&B: *Ephedra* Sporophyte.

A - The plant is much branched and some species are low height shrubs. B - The articulated stem with its nodes and internodes resembles *Equisetum*.

small, woody and branched shrubs not exceeding 3 meters in height and when the species is a climber, it is fistular, and straggling stem. In the axil of leaves arise two to four branches which further branch. These branches which arise from the lower nodes, are closely placed and give plant a bushy appearance **(Fig. 4.13A&B).** Due to their arrangement and structure they resemble those of *Equisetum*. The plant remains attached to the ground by long tap root system.

Anatomy

Stem

- Stem is ribbed and therefore, the outline is not circular but has ridges and furrows that unlike *Equisetum* are not well demarcated.
- Epidermis has outer cuticle and is interrupted by sunken, haplocheilic stomata which remain restricted to the furrows.
- Hypodermis consists of a group of fibrous cells.
- Since stem has taken over the function of the leaves, it has become photosynthetic with outer palisade and inner spongy parenchyma (**Fig. 4.14; Plate IV.2A&B).**
- Endodermis and pericycle are two layers present outside the stele, latter being generally indistinct.
- Stellar region has two pairs of small (for leaves) and four large (for stem) vascular bundles arranged in a ring; bundles are conjoint, collateral and open. Thus, each leaf is provided with a pair of leaf trace bundles.
- The medullary rays are uniseriate.
- Pith is the innermost parenchymatous tissue well defined in the nodal region where it is lignified to form a peridermal diaphragm **(Plate IV.2B)** but is not so in the internodal region.

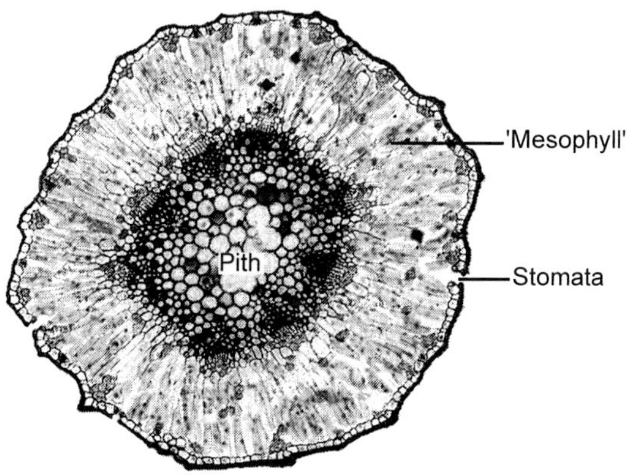

Figure 4.14: *Ephedra*T.S. Stem shows the undulated outline, well developed 'mesophyll' and stomata. Also seen are the vessels in the xylem.

Secondary Growth

The older stem becomes thickened due to the activity of cambium; the medullary rays appear broad and lignified. Secondary tissues are formed in a normal manner from inter and intra-fascicular cambium.

Tracheids have a single row of bordered pits both on the lateral and radial walls. Also seen are vessels with foraminate perforation plates on end walls. The spring wood is rich in such vessels which decrease in number from the autumn wood. Transitional stages in formation of vessels from tracheids are also seen. There are no resin ducts but calcium oxalate crystals may be seen. Secondary cortex arises from phellogen.

Leaf

- Leaves which arise at the node are scaly, highly reduced and form a sheath around the node **(Fig. 4.15A&B).**
- It is technically non-photosynthetic with a simple structure.
- The epidermis which is interrupted by sunken stomata, is followed by single layered palisade and two closed collateral vascular bundles are seen.

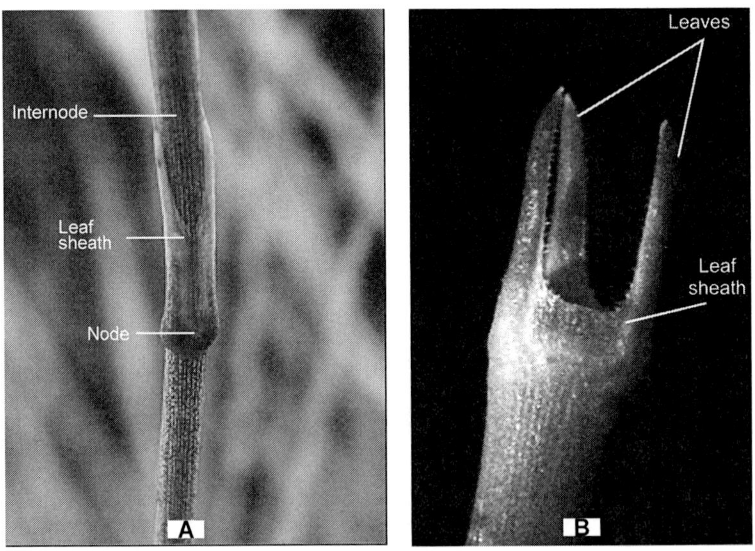

Figure 4.15 A&B: *Ephedra* stem with internode and node.

A - Nodal region bearing leaves. Note the leaf sheath extending to the internode.
B - Node magnified to show small scaly leaves which are fused at the base forming a sheath and remain free at the tip.

Reproduction

Vegetative Reproduction

Certain species like *Ephedra nevadensis* propagate vegetatively by shoots which after being dropped on the ground get buried and form roots. After some more development, they form lateral branches. Even the rhizomes in some species give rise to new plants.

Sexual Reproduction

The dominant plant is a dioecious (rarely monoecious) sporophyte with compound and unisexual strobili borne singly or in pairs or whorls at the branch nodes **(Fig. 4.16).** In most species, staminate and ovulate strobili start appearing in December and pollen is shed in about April followed by fertilization that occurs after a few hours only.

The distribution of male and female *Ephedra* plants is not random; individuals on dry slopes are over-represented by males, whereas those growing on run-on surfaces are four times as likely to be females as males. The genetic basis for sex differentiation in *Ephedra* is not known, but the spatial arrangement of males and females functions to maximize reproductive output. It places males where their pollen can be easily wind-dispersed early in the season and females where they are more likely to have resources later in the season to ripen a seed crop (Meyer, 2008).

Ephedra plants do not 'flower' every year; their reproductive pattern could be described as mast fruiting, where most individuals in the population flower synchronously in a year with ample rainfall, and large quantities of seeds are produced. The population does not flower again for several years, whether or not a high-rainfall year occurs.

Figure 4.16: *Ephedra*. Portion of the reproductive axis with a cluster of branches bearing strobili at the nodes.

Male Strobilus

- The male strobili are compound structures arising in clusters from the nodes of the branches **(Fig. 4.17A; Plate IV.3).**
- Each strobilus develops in the axil of a scaly leaf, more commonly referred to as a bract.
- Male strobilus is small with a short-stalk/axis bearing four-sixteen pairs of thick bracts which are opposite and decussate **(Fig. 4.17B).**
- In the axil of each bract, excepting the lowermost one or two pair/s which are sterile, is present a staminate or male flower **(Plate IV.4)**.

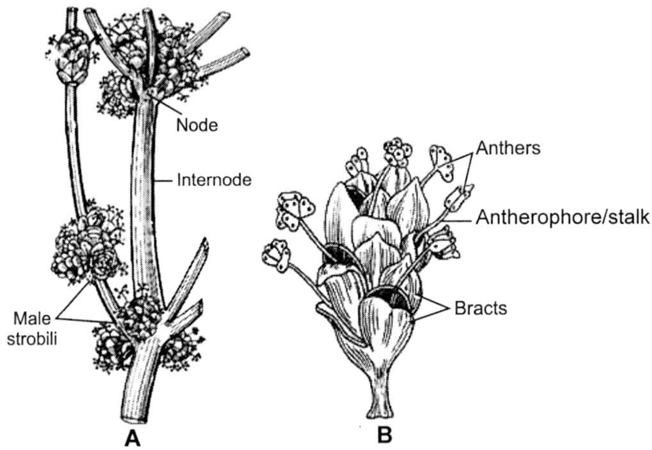

Figure 4.17 A&B: *Ephedra*. Male reproductive structures.

A - Portion of the male plant showing cluster of male strobili borne at a single node.
B - One strobilus is enlarged to show paired opposite decussate bracts in axil of which male flowers are borne.

Male Flower

- Each flower has a small axis or microsporangiophore/antherophore bearing at its tip about eight anthers or microsporangia (**Fig. 4.18A&B; Plate IV.5).**
- The entire structure is invested by two connate bracteoles or the perianth **(Plate IV.6A).**
- The microsporangia have either two or three loculi each opening by a terminal slit at the time of dehiscence. They contain pollen grains and remain protected by sterile structures **(Plate IV. 6A&B)**.
- The haploid microspores or the pollen grains elliptic in shape, with ribbed wall, are dispersed by wind **(Fig. 4.18C).**

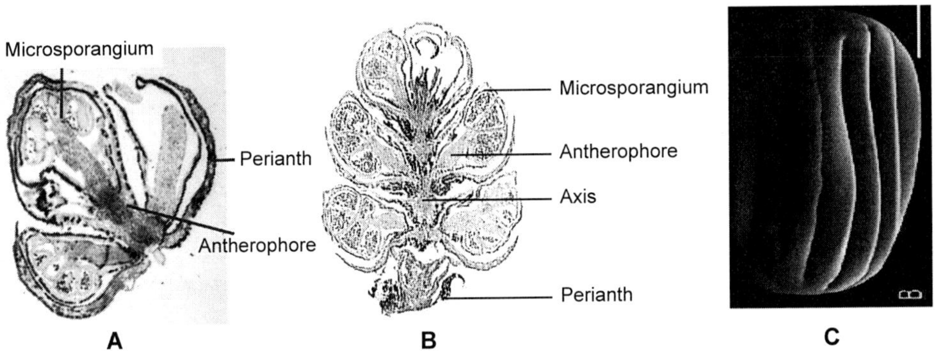

Figure 4.18 A-C: *Ephedra* Male strobilus.

A - T.S. Male strobilus to show perianth and microsporangia at the tip of the antherophore or microsporangiophore. B - L.S. Shows the central axis around which the microsporangia are arranged. The wall of the sporangium and microspores are distinct. C - Microspores are ribbed.

Female Strobilus

- The female branch bears ovulate strobili on the nodes **(Plate IV.7A)**.
- The number of female strobili is much less than the male counterpart on the node.
- Even the number of female flowers is much less than the male flowers per strobilus **(Fig. 4.19A-D)**.
- It is sub-sessile to sessile, ovulate strobilus with a central axis and scaly bracts about two to four in pair, arranged in opposite and decussate manner.
- The lowermost pair is fused to form a cupule **(Plate IV. 7B)**.
- Besides being less in number than the male flower, the bracts in female may be coloured or dry or winged.
- Each strobilus ends in a group of flowers, a few in number; *e.g.*, in *E. foliata*, the uppermost pair of bracts is the largest and encloses two flowers where usually one aborts while the other which gives rise to a seed upon fertilization, becomes terminal in position **(Fig. 4.20A&B; Plate IV. 8A&B)**.

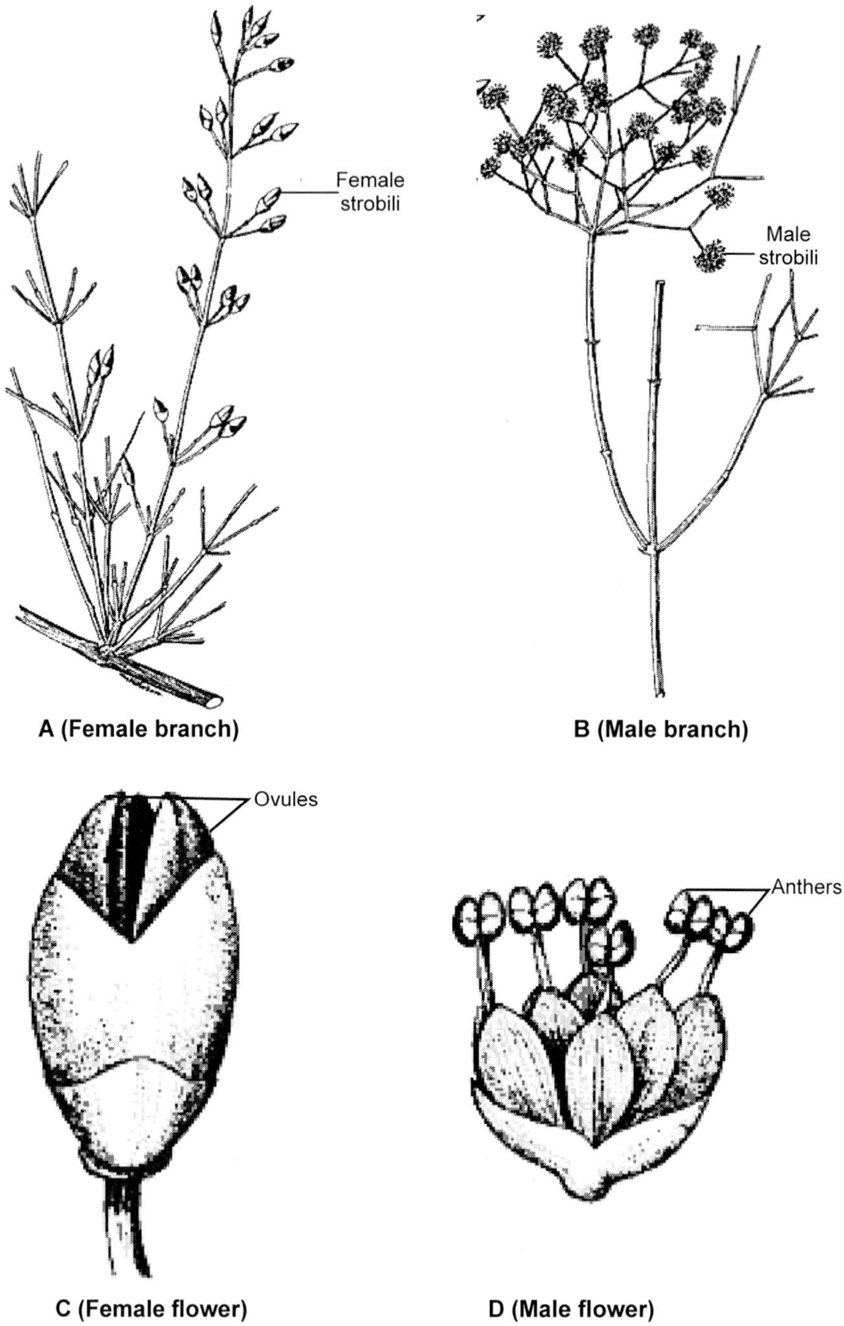

Figure 4.19 A-D: *Ephedra*. Comparison between the female and male structures.
A - Female branch. B - Male branch. C - Female flower. D - Male flower. Note the number of strobili and female flowers is much less than the male flowers.

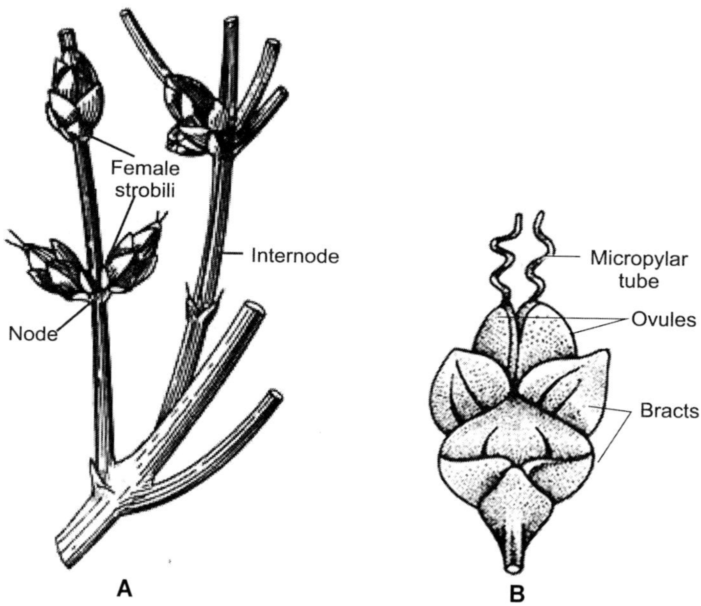

Figure 4.20 A&B: *Ephedra* Ovulate axis.

A - Note the number of bracts forming the female strobilus is less. B - Pair of female flower is present in the terminal bract pair.

Ovule

- The female flower is represented by a short-stalked terminal megasporangium or an ovule. The ovule consists of a nucellus enclosed with two integuments. The inner corresponds to true ovular integument while the outer corresponds to the perianth. They are provided by two and four vascular bundles respectively.
- Inner integument fuses with nucellus in lower half but its upper half remains free and is prolonged into a thin-walled micropylar tube which is twisted, opening at its tip into a tongue-like extension, the tent pole **(Fig.4.20B; Plate IV. 9A&B).**
- A conspicuous pollen chamber is formed in which the pollen grains come to rest before germination.
- The pollination drop is exuded by means of which pollen is drawn down the micropylar tube into the pollen chamber.

Megasporogenesis

Archesporial cell which gets differentiated in the nucellus gives rise to a megaspore mother cell. This cell divides meiotically to produce a T-shaped or linear tetrad of megaspores; lowest one being functional. The development of the female gametophyte is therefore, monosporic.

Gametophytes

Male Gametophyte: Pollen Grain

Pollen grain has (after Frernando *et al.*,2010) a prothallial cell, a prothallial nucleus, and an antheridial cell in the beginning **(Fig. 4.12E)**. As the development proceeds, the antheridial cell and a tube cell are formed. The former gives rise to two sperm nuclei. At the time of anther dehiscence and pollen dispersal, the male gametophyte is a two celled and three nucleate structure **(Fig. 4.21A-D)**.

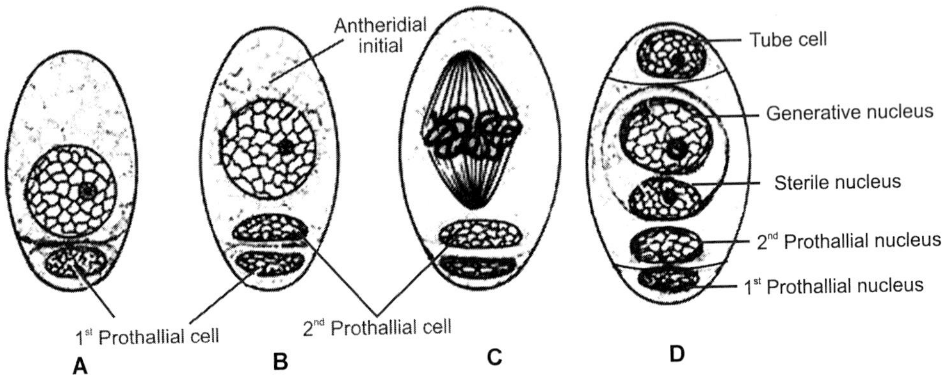

Figure 4.21 A-D: *Ephedra*. Development of the male gametophyte. The pollen grain is shed at two-celled and three nuclei stage.

Upon pollination, generative nucleus divides unequally into male or sperm nuclei with equal volume. The two male nuclei and tube nucleus pass into the pollen tube while the two prothallial entities degenerate.

Female Gametophyte

Female gametophyte develops from the chalazal functional megaspore of the tetrad formed by meiotic division of the megaspore mother cell. During development of female gametophyte, the nucleus of functional megaspore undergoes free nuclear divisions. In *Ephedra trifurca* about 256 nuclei and 500 in *E. foliata* are formed.

This first phase of growth is followed by a period of centripetal wall formation or alveolarization. This gives rise to a multicellular female gametophyte similar to other gymnosperms.

Following cellularization, archegonia are initiated from an initial at the micropylar end of the gametophyte. The female gametophyte shows distinct regions where the micropylar one is fertile region in which the archegonial initials differentiate and the lower is the sterile haustorial region **(Fig. 4.22A)**. Each archegonium initial divides to form a primary neck cell which forms a multitiered neck and a central cell. Archegonia remain at this stage till pollination (Moussel 1977, Friedmann 1990). Later, the central cell nucleus divides and a mature archegonium with a long multilayered neck and a central cell containing a ventral canal nucleus and an egg nucleus is formed. The long columnar neck appears similar to the transmitting tissue in the style of angiosperms. Subsequently, the egg nucleus migrates in a chalazal direction towards the centre of archegonia where it becomes immersed in a column of cytoplasm rich in both mitochondria and plastids. In contrast, the micropylar region of the egg cytoplasm is vacuolate and largely devoid of most cytoplasmic organelles **(Fig. 4.22B)**

At the time of pollination the ovule contains one to six archegonia, each of which is surrounded by a multi-layered jacket (Maheshwari and Sanwal, 1963). Jacket cells are bi- to multi-nucleate with dense cytoplasm. A pollen chamber develops in the ovular apex where pollen germinate. *Ephedra* has the deepest pollen chamber amongst gymnosperms which closes upon pollination.

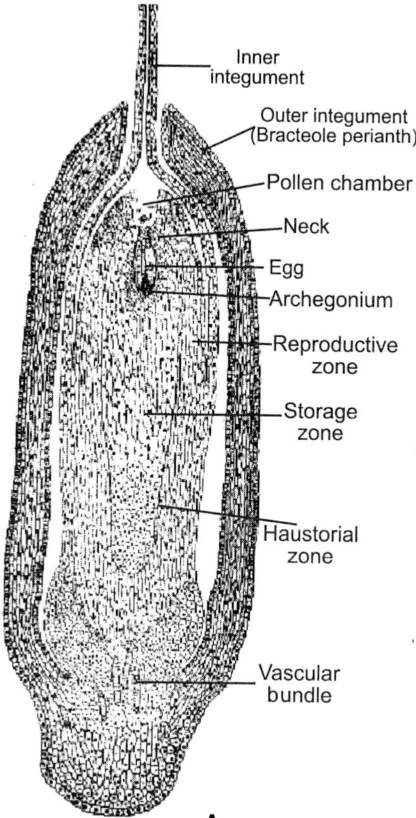

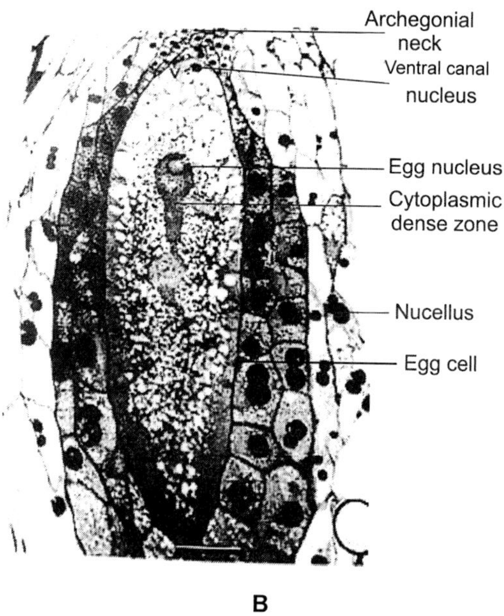

Figure 4.22 A&B: ***Ephedra*****. L.S. Ovule.**

A - To show various tissues of the ovule. Tent pole formed by the inner integument, well differentiated female gametophyte and a distinct pollen chamber are features of a mature ovule. B - Mature archegonium with two potential nuclei to fuse with sperms – the ventral canal nucleus and the egg nucleus.

The walls of the jacket cells breakdown at the time of fertilization and some nuclei of the jacket have been reported to move into the central cell just prior to fertilization.

Thus, in addition to egg nucleus and ventral canal nucleus, are also present in the archegonium other, female nuclei of jacket origin. According to Land (1907), migratory jacket nuclei might participate in supernumerary fertilization events. This view has now been rejected on the basis of microspectrophotometric analysis of DNA content of the nuclei (Russell and Dumas, 1992).

Pollination

At anthesis a sugary pollination drop is formed by the nucellus and secreted through a micropylar tube to the outside of the ovule (Rydin *et al.*, 2010). The pollination drop functions, like in most gymnosperms (Owens *et al.*, 1998), as a receptive site and as transportation and germination medium for pollen (Gelbart and von Aderkas, 2002). In *Ephedra* only female strobili produce pollination drops because the male strobili do not have ovules unlike those in *Gnetum*. However, in *E. foeminea*, (Rydin and Korall, 2009; Thureborn and Rydin, 2015), there are sterile female organs present in the male strobili that produce sugary pollination drops even though they cannot produce viable seeds. Both *Gnetum* and *Welwitschia*, the two other members of the group, are described as insect pollinated (Kato and Inoue, 1994; Kato *et al.*, 1995; Wetschnig and Depish, 1999; Gong *et al.*, 2015) and are known to produce sugary pollination drops through sterile female organs in the male strobilus (Endress, 1996). For successful pollination with insects as vector to occur, there are three main aspects that need to be fulfilled 1) attachment of the pollen at the vector 2) transportation to the receptive site and 3) detachment of the pollen at the receptive site. In dioecious species, the possibility to attract insects first to the male plant and then to the female plant is, thus, essential in order for successful insect pollination to occur.

Though most species of *Ephedra* are wind pollinated but *E. foeminea* appears to be adapted for insect pollination (Bolinder *et al.*, 2016).

Fertilization

Elapsed time between pollination and fertilization is extremely brief, 10 hrs in *E. trifurca* (Land, 1907) and 16 hrs in *E. distachya* (Moussel, 1977).

The central cell nucleus divides to produce a ventral canal nucleus and an egg nucleus. There is no evidence of wall formation between the two (Friedman, 1990).

The ventral canal nucleus remains *in situ* at extreme micropylar end of the egg cytoplasm and persists through the time of fertilization (Friedmann, 1991). Egg nucleus migrates in the chalazal direction toward the centre of the archegonium (Moussel, 1977). The micropylar region is more vacuolated at this stage.

Fertilization of the egg nucleus with one of the two sperms leads to the zygote formation. Herzfield (1922), reported that the second sperm often fuses with the ventral canal nucleus in *E. campylopoda*. While the first fertilization leads to zygote formation from which upon further development eight proembryos arise, the product of second fertilization between second sperm and ventral canal nucleus, soon disintegrates.

Post Fertilization Changes

Coincident with fertilization and development of the proembryos, the gametophyte is stimulated to further activity. A single plate of the outermost layer of cells of the gametophyte, forming the bottom of the pollen chamber and extending outward a short distance under the tip of the nucellus, becomes meristematic. By repeated periclinal and a few anticlinal divisions it forms a plug which effectually closes the pollen chamber and receives the backward thrust of the elongating cells of the gametophyte and later of the embryo. This plug persists until absorbed by the maturing embryo. The bulk of the gametophyte is materially increased. The cells near the archegonia first elongate, followed successively by groups of cells extending across the gametophyte. The elongated cells after having lost their contents are folded back and forth in the micropylar region, being prevented from escaping through the pollen chamber by the plug. This elongation of groups of cells, proceeding successively from the micropylar to the antipodal region, serves to keep an abundance of food constantly near the embryo. As the seed matures, the integuments and nucellus become gorged with food.

Embryogeny

Although, the embryogeny of only a few species of *Ephedra* has been worked out, various accounts are more or less similar (Lehmann-Baerts, 1967). The zygote divides *in situ* and the two nuclei move apart, sometimes occupying the two poles of the zygote. Two more mitoses follow to produce eight nuclei. Each nucleus is surrounded by a densely staining cytoplasmic sheath with radiating strands. At the eight nucleate stage an independent cell wall develops around each nucleus resulting in eight units that later develop into eight proembryos, also referred to as proembryonal masses. Each produces a tubular outgrowth. The proembryo nucleus divides and moves into the tube. A transverse wall is laid down giving rise to an embryonal cell towards the tip and a distal suspensor cell. The suspensor cell quickly elongates, thrusts the embryo initial deeply downward into the gametophyte, and remains turgid until the secondary suspensor appears, when it collapses. The embryonal cell comes to lie deeper in the gametophyte, it divides and distal cell elongates to form secondary suspensor cells. The secondary suspensor is added by successive elongations of basal cells of the embryo and finally merges into root cap (**Fig. 4.23A-G**). The embryo divides and redivides and differentiates a dermatogen. *Ephedra* shows both cleavage and simple polyembryony (Kozlowski, 1972).

As the seed matures, the integuments and nucellus become gorged with food. The epidermal cells, especially those of the nucellus are filled with starch and become so turgid that they are frequently torn loose from the layer beneath. In some habitats, the young plants are seen to escape from strobili still attached to the stem. This shows that under favourable conditions of growth, the period of

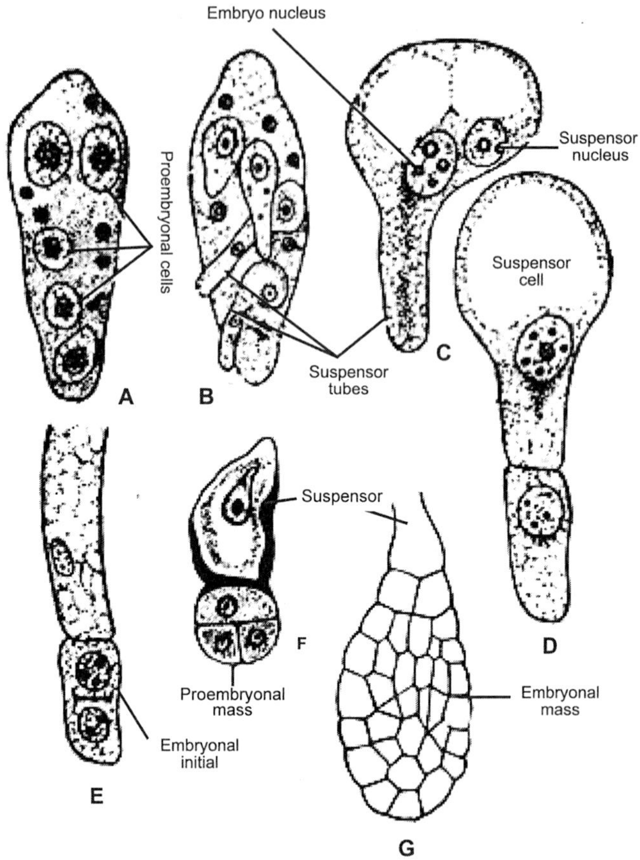

Figure 4.23 A-G: *Ephedra*. Embryogeny.

A - Number of proembryonal masses are formed as the zygote undergoes development. B - Later proembryonal tubes develop. Further development results in formation of embryonal mass from which embryo arises. C-G - The suspensor cells elongate pushing developing embryo into the gametophyte.

rest of seed is a brief one. The time from setting of the strobili to the germination of the seed is, therefore approximately six months (Pearson, 2010).

Seed

The seed contains a dicotyledonous embryo. It is situated at the tip of the elongated suspensor and remains embedded within the tissue of the female gametophyte. The remnants of nucellus are seen in the form of a disorganised sheath of cells **(Fig. 4.24).**

The seed coat consists of two separate layers derived from two layers of the envelope. At maturity the subtending adjacent bracts of the strobilus in *Ephedra foliata* become thick and fleshy and form an additional covering over the seed. Thus, the seed remains covered by three envelopes.

The seeds are borne singly or in pairs in the axils of the female bracts. The inner bracts or scales are modified to enclose the seed and form integuments that mimic the angiosperm pericarp.

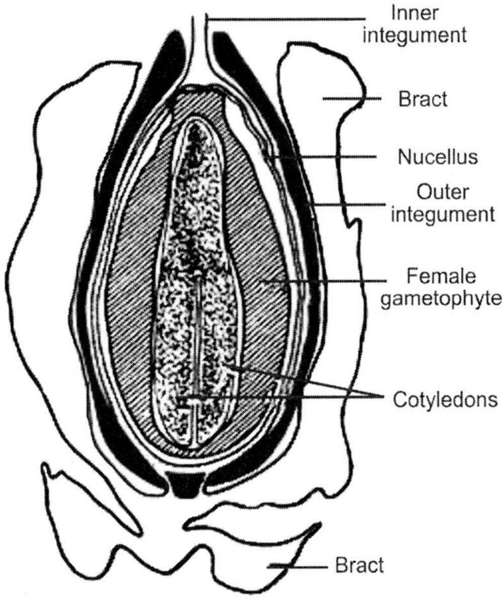

Figure 4.24: *Ephedra* a mature seed shows two cotyledons and endosperm. Note the presence of nucellar remnants and the bracts.

North American *Ephedra* species fall into two groups characterized by differences in seed size and dispersal ecology. The large-seeded species (for example, green and Nevada Mormon teas) are dispersed by scatter-hoarding rodents such as kangaroo rats (*Dipodomys* spp.), which deposit them in shallowly buried caches and later return to eat most of the seeds or sprouts. The strobilus bracts in these species are small. In small-seeded species (for example, Torrey Mormon tea) the outer strobili scales are large and membranous, and the intact strobili are often seen windrowed at some distance from adult plants. The seeds are apparently wind-dispersed, as they have long, awn-like points that probably make them unattractive to rodents. Strobili with seeds intact may remain on the surface for many months.

**In 1997, in response to mounting concern over serious side effects of Ephedra (drug), the FDA proposed a ban on products containing 8 mg or more of ephedrine alkaloids and stricter labeling of low-dose Ephedra supplements. The FDA also proposed that Ephedra labels be required to disclose the health risks such as heart attacks, strokes and death.*

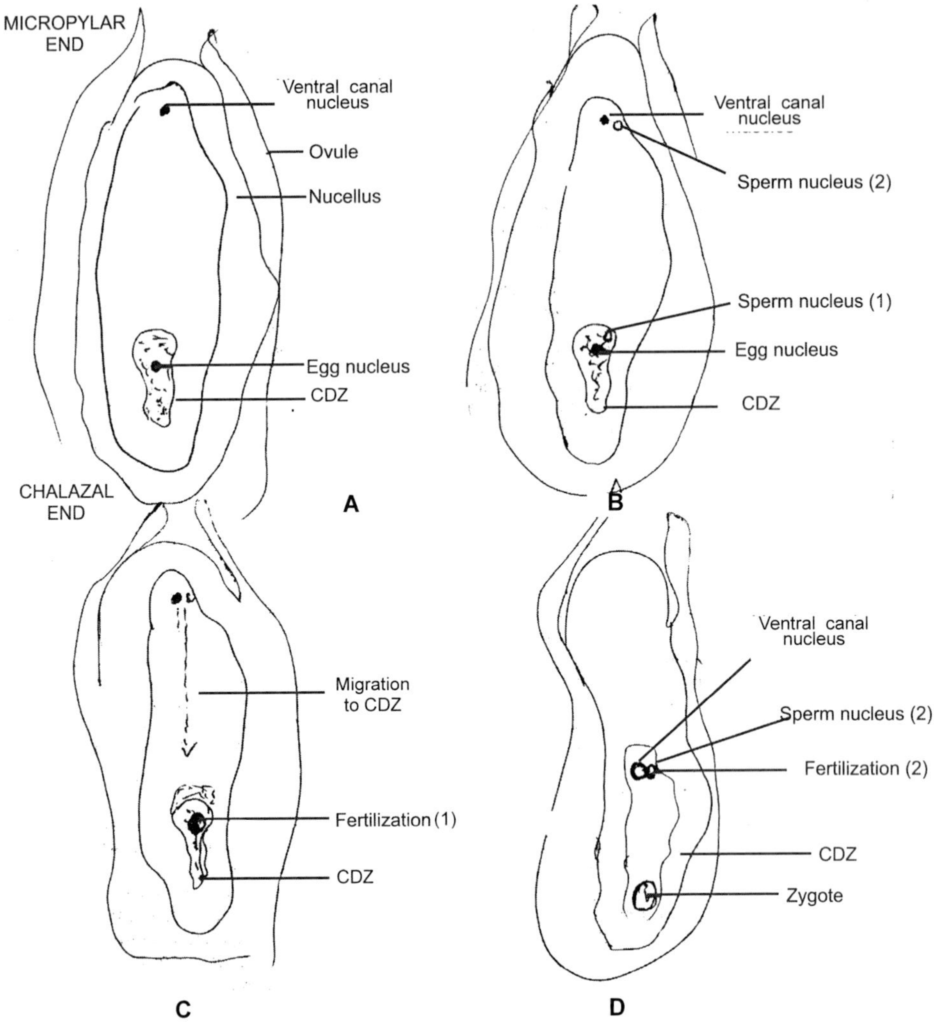

Figure 4.25 A-D: *Ephedra*. Stages in double fertilization.

A - Ventral canal nucleus remains *in situ* while egg nucleus migrates to chalazal end and stays in cytoplasmic dense zone (CDZ). B - Out of two sperm nuclei, one moves to CDZ and fuses with the egg nucleus resulting in first fertilization which would produce a zygote. C - Once the fusion of egg and sperm nuclei is over, the second sperm starts its journey along with venter canal nucleus to the CDZ. D - Reaching there, the two nuclei fuse resulting in second fertilization, the product formed gets degenerated soon after formation.

Salient Features of Sexual Reproduction

The zygotic nucleus undergoes free nuclear divisions resulting in eight free nuclei which remain evenly distributed throughout the cytoplasm. Each nucleus then develops into a cell, and each of such cell functions as a potential pro embryo. Khan (1943) observed that six out of these eight nuclei get surrounded by globular cells and function as six potential embryos. However, all eight proembryos may be distinct. Such a stage represents polyembryony without any cleavage.

Herzfield (1922), reported that the second sperm often fuses with the ventral canal nucleus in *E. campylopoda*. Maheshwari (1935), suggested that in exceptional cases the ventral canal nucleus may behave like an egg nucleus if displaced from its normal apical portion within the egg cytoplasm to a more central position.

In *E. nevadensis* and *E. trifurca*, the first sperm nucleus immediately migrates to the egg nucleus. As the contact is made between the two gamete nuclei, the egg nucleus becomes invaginated. This pair of nuclei then migrates to the chalazal end of the egg cytoplasm. When the migration is over, the two nuclei are no more discernable and they fuse to yield a zygote.

Shortly after this contact between first sperm nucleus and egg nucleus, the ventral canal nucleus and the second sperm nucleus migrate towards the chalazal end and the pair rests till the fertilization between egg and first sperm nucleus is accomplished. Following this, the fusion between second sperm and ventral canal nucleus occurs.

The fate of the second fertilization product has been studied in *E. trifurca* (Friedman, 1992). At the end of this intra-archegonial double fertilization, the two diploid nuclei reside within the common cytoplasm. The fusion product of second fertilization may soon get disintegrated and the embryo is initiated from the fertilization between egg and the first sperm nucleus.

Economic Importance

"The stems contain caffeine and ephedrine (a drug that acts like adrenalin/ epinephrin). "Mormon tea (*Ephedra nevadensis*) is made from the dried branches when boiled in water. People use it as a beverage for stimulation of heart. Unlike other plants, Mormon tea does not contain ephedrine, an unsafe stimulant. Decoctions of stems treat a variety of ailments including diabetes, pneumonia and sexually transmitted diseases such as syphilis and gonorrhea. It is also used for colds, kidney disorders, and as a 'spring' tonic." A Chinese species is the source of ma huang, a tea so potent that it has caused deaths from overstimulation of the heart."

"The tannins in Mormon tea have a drying (astringent) effect and can reduce body secretions such as mucus. However, there are reports of side effects such as stomach complaints, kidney and liver damage, nose or throat cancer, increased urination, and constipation."

Plants of the *Ephedra* have traditionally been used by indigenous people for a variety of medicinal purposes, including treatment of asthma, hay fever, and common cold. The alkaloids ephedrine and pseudoephedrine are active constituents of *E. sinica* and other members of the genus. These compounds are sympathomimetics, with stimulant and decongestant qualities and are related chemically to the amphetamines.

Do You Know?

Bracts of female cones of extant gymnosperm *Ephedra* (Joint fir) are either colorful and fleshy (section *Ephedra*), or dry-winged and membranous (section *Alatae*), or dry and coriaceous (section *Asarca*), which have played a crucial role in long-distance seed dispersal that is responsible for a wide distribution of the genus in semi-arid and arid areas of Eurasia, North Africa, North America, and South America. Recent molecular systematic studies on *Ephedra* have suggested that the fleshy bracts in character evolution may be plesiomorphic relative to the dry, membranous and coriaceous bracts. However, little is known about when the fleshy bracts of *Ephedra* have made their debut in the geological past.

Test Your Knowledge

Q 1. List out the xerophytic characters of *Ephedra.*

Q 2. Describe the male and female strobili in *Ephedra.*

Q 3. Draw L.S. - male flower, female flower, seed.

Q 4. Define: tent pole, foraminate pitting, prothallial cells, double fertilization

Q 5. Discuss double fertilization in *Ephedra.*

Chapter 5

Subclass IV Pinidae

Pinidae is divided into three orders Pinales, Araucariales and Cupressales. The subclass which comprises conifers, is equivalent to the division Pinophyta and class Pinopsida.

The two orders - Araucariales and Cupressales have Auraucriaceae and Podocarpaceae as two families in the former and Sciadopityaceae, Cupressaceae and Taxaceae as three families in the latter, while Pinales the third order is monotypic with Pinaceae as its family.

The Success Story of Conifers

The conifers outdate all flowering trees by a vast span of time. The first conifers discovered dates back to the late Carboniferous period (about 300my), while the first fossils of angiosperms, which probably originated earlier in the Mesozoic, appeared in the Cretaceous (about 15my).

The first known conifer trees resemble modern *Araucaria* trees. They perhaps evolved from an ancestral gymnosperm tree which could live in warm swampy conditions – as watery habitat was a prerequisite since these trees could not cope well with dry conditions. However, that appeared later in evolution when the evergreen cone bearing trees could survive in drier conditions. The early conifers had immense competitive advantages which were critical to survival as the great warm swamps of the Carboniferous dried out. The continents which had once been separated by shallow oceans and seas, joined together into warm to hot dry supercontinent of Permian Pangaea. In the arid deserts and mountains, the conifers were among the only plants which could survive. This ability to live through any condition helped conifers get through the greatest mass extinction in lifes' history – THE PERMIAN-TRIASSIC (P-Tr) extinction event (known to palaeontologists as 'the great dying'). Thereafter, throughout the Mesozoic, conifers were the dominant land plants (along with cycads and ginkgos which had evolved at about the same time). The Mesozoic saw the greatest diversity of conifers ever – THE AGE OF DINOSAURS could just as well be called — THE AGE OF CONIFERS.

Although flowering plants rapidly came to prominence towards the end of Cretaceous and have since become the most diverse plants, todays conifers are not in any way anachronisms (idea that exists out of its time in history). They still outcompete the flowering trees in cold areas and in dry areas. Conifers dominate the boreal forests of Asia, Europe and North America.

Conifers are probably the most commercially important trees for pulp and timber industry.

The conifers are nearly everywhere and are a symbol of life surviving through the winter to come back stronger. Conifers tend to have shallow roots and are subject to being knocked down but, the pyramidal shape reduces wind resistance and helps keep tree standing upright. The shape of leaves also helps blowing ice crystals and allows lower branches with more light. Branches have an open area between, allowing wind to pass through.

Some of the reproductive pathways also helped in conifer spread. Instead of relying on free swimming gametes, the motile spermatozoids for fertilization and huge seeds (*Ginkgo* and cycads), the newly evolved conifers used wind to carry 'clouds' of pollen through the air and were capable of producing many tiny seeds which could survive drying out.

In conifers a variety of pollen reception mechanisms evolved and these finalized the conversion to the water independence of spores for movement and fertilization. The inverted ovule's orientation is related to the type of pollen present. Seeds in addition to mycorrhizal associations and xylem with tracheids in roots, are considered the major evolutionary factors allowing conifers to dominate large land areas especially those with stressful or sub-optimal environments.

Most conifers show trio of traits that provide an ancient function of pollination. Many pine and spruce species still exhibit attributes like pollen grains with sacci that are buoyant, downward falling ovules and well-timed emission of a drop of liquid (pollination drop) from the ovules. These traits are gradually disappearing in extant conifers. This trio of traits began to break in many species. Some clades lost buoyant saccate pollen, while others dropped the droplet and some re-oriented their ovules. According to researchers when the mechanism broke down for conifers, they became free to put the traits to other uses or evolve entirely new ones. Many species have some but not all three traits *e.g.*, Junipers still emit liquid drops to capture pollen from wind but saccate pollen is absent. Hemlocks evolved completely with pollen that can land anywhere on the cone and grow tubes towards it. Thus, pollination by floatation is interesting case of evolution.

Order Pinales

This order was earlier known as Coniferales. The distinguishing characteristic is the reproductive structure known as cone, produced by all Pinales, both the male and female plants. The *conifer assemblage has been reconsidered in light of molecular studies, which have shown Pinaceae to be sister to all other conifers (a group comprised of two sister clades, Araucariaceae + Podocarpaceae and Cupressaceae + Sciadopityaceae + Taxaceae) (for details see Missouri Botanical Garden, 2011).

Since the collective name conifers has been used synonymously with pines, it would be worthwhile if characters of 'Conifers' are listed for clarity. The features are:

- Conifers were abundant in the Mesozoic with many extinct forms occurring well into the Cretaceous.
- They typically have straight trunks with horizontal branches varying more or less regularly in length from bottom to top, so that the trees are conical in outline.
- Nearly all plants are trees (rarely shrubs) with needle-shaped or scale-like leaves which are provided with one or two veins, although in some genera the leaves are broad and contain many vascular bundles (*e.g., Agathis*).
- The plants bear dimorphic shoots – long and short or dwarf shoots.
- With the exception of a few deciduous taxa i.e.; *Larix, Taxodium, Metasequoia, Glyptostrobus*, and *Pseudolarix*, all are evergreen, and their foliage functions for more than a single growing season. In some high latitude populations of *Pinus longaeva*, the foliage may remain on the plant up to 40 years!
- The oldest living plants are found among the conifers, a bristlecone pine (*Pinus longaeva*) in eastern Nevada is estimated to be a little less than 5000 years old.
- Wood is massive with small pith (pycnoxylic) and is characterized by circular bordered pits; the primary vascular system is a eustele with endarch xylem.
- Most members are monoecious with male and female cones on same plant, though on different branches.
- They are characterized by having staminate or pollen-producing cones and most of them also bear compound ovulate or seed-producing cones.

*Conifers' therefore, comprise two sister clades with above mentioned five families and Pinaceae.

*According to Christenhusz *et al.* (2011), the term 'Coniferales' has been frequently used but it is not based on an existing genus. The use of names based on 'Conifer-' (*e.g.,* Coniferopsida, Coniferidae, Coniferales *etc.*) should therefore, be avoided. (Available from: https://www.researchgate.net/publication/228691404_A_new_classification_and_linear_sequence_of_extant_gymnosperms).

- Ovules are borne on the adaxial (upper) surface of the ovuliferous scales, and the entire unit is termed the bract–seed-scale complex.
- Mature seeds are usually rather small with a hard seed coat.
- Almost all conifer roots form arbuscular mycorrhiza, but the species in the Pinaceae form ectomycorrhizas (Brundrett, 2008).
- Species in the Pinaceae are almost entirely confined to the northern hemisphere.

Family Pinaceae

The family with 11 genera, ca 225 species, is well distributed in temperate to tropical Eurasia, Sumatra, Philippines, North America, south to Nicaragua, and West Indies. The phylogenetic tree published by Liston *et al.* (2003) has been used to create sequence of eleven genera - *Cedrus, Pinus, Cathaya, Picea, Pseudotsuga, Larix, Pseudolarix, Tsuga, Nothotsuga, Keteleeria* and *Abies*. (https://www.sciencedirect.com/topics/agricultural-and-biological-sciences/pinales).

- The Pinaceae (pine family) members are trees or shrubs, including many of the well-known conifers of commercial importance such as cedars, firs, hemlocks, larches, pines and spruces.
- Pinaceae are supported as monophyletic by their protein-type sieve cell plastids, pattern of proembryogeny, and lack of bioflavonoids.
- It is the second largest family (after Cupressaceae) in geographical range, found in most of the Northern Hemisphere, with the majority of the species in temperate climates, but ranging from subarctic to tropical.
- The family often forms the dominant component of boreal, coastal, and montane forests. One species, *Pinus merkusii*, grows just south of the equator in south east Asia. Major centres of diversity are found in the mountains of south west China, Mexico, central Japan, and California.

Most of the genera are well adapted for cutting down water loss. Adaptations exhibited by the plants for growth in physiologically dry soil are:

- The conical or spire-shaped needles can adapt tress to the cold and the drought of winter and short-growing season.
- This type of tree conserves energy by not regrowing its leaves during spring and can shed snow more easily.
- The narrowness of needle leaf reduces surface area through which water may be lost (transpired), especially during winter when the frozen ground prevents plants from replenishing their water supply.
- Evergreen habit wherein retention of foliage allows plants to photosynthesize as soon as temperatures permit in spring, rather than having to waste resources in the short-growing season, merely growing leaves.

- The dark green region of spruce and fir needles helps the foliage absorb maximum heat from the sun and begin photosynthesis as early as possible.

Pinus

Distribution

Pinus is the Roman name for pine. The genus is the largest in the family, with ca 114 known species and is one of the oldest extant conifer genera. The oldest pine known is *Pinus mundayi*, with two needles described from fossil charcoal of the lower Cretaceous (ca. 133–140 Ma) Chaswood Formation of Nova Scotia, Canada (Lang *et al.*, 2016). The charcoal occurrence of this fossil is interesting, as it suggests that this extremely fire-adapted genus may have had an association with fire since the beginning. A comparable ancient fossil pine is *Pinus yorkshirensis*, documented from a fossil cone preserved in the Wealden Formation in Yorkshire, UK. These sediments are about 130 my old (Ryberg *et al.*, 2012).

Earlier two subgenera were recognized and the typical pines were placed in subgenus *Pinus* and the white pines in subgenus *Strobus*. The subgenera were chiefly distinguished on the basis of several criteria (Richardson, 1998).

When categorized, subgenus *Pinus* L., also called the hard pines had about 70 species. Cone scales had a dorsal, mucronate umbo and a sealing band adjacent to the apophysis where the scales met on the closed cone. There were two vascular bundles per needle. Each fascicle had two to six needles; with stomata more or less equally distributed on all surfaces, resin ducts were variable; sheath persistent except in *P. leiophylla* and *P. lumholtzii*; the fascicle bases had decurrent pulvini.

Subgenus *Strobus* L. also called white or soft pines, on the other hand included about 44 species. Cone scales without a sealing band and seed wing were articulate to strongly adnate. Only one vascular bundle per needle was seen and each fascicle had one to five needles, stomata were reported from all or mostly on inner faces, resin ducts were medial or external; the fascicle sheath was deciduous except in *P. nelsonii*, and the fascicle bases had non-decurrent pulvini.

Though the present classification does not accept these subgenera but researches indicate that *Pinus* is one conifer genus that, despite its antiquity, shows ample evidence of ongoing speciation (www.conifers.org/pi/pinus).

Pines are not part of super rich forest assemblages. In the north temperate zone where it is widely distributed, it is rivalled by *Quercus*. The pines extend from polar region to the tropics, crossing the Equator. They dominate the natural vegetation in many regions (Critchfield and Little, 1966).

Six species of pines are indigenous to India, of which *Pinus roxburghii* popularly known as 'chir', is very common and grows luxuriantly in western Himalayas at a height of 460 to 1500 m. *Pinus wallichiana* (blue or Bhutan pine) commercially known as Kail is also found in western Himalayas at a height of 1500 to 3000 m. *Pinus insularis* is restricted to eastern Himalayas at a height of 800 to 2000 m and is commonly known as 'Khasi' pine. *Pinus gerardiana* commonly known as 'chilgoza' pine is known for its edible seeds. It occurs at an elevation of 1830-3600 m. *Pinus armandii* is restricted to north eastern Himalayas while *Pinus merkusii* (Tenasserim pine) is found in hill slopes of Andaman and Nicobar islands.

Sporophyte

- The main plant is a tree or a shrub, aromatic, and evergreen.
- The canopy or the crown is usually conic when young, often rounded or flat-topped with age.
- Bark of older stems is variously furrowed and plated, where plates and/ or ridges are layered or scaly.
- Branches usually arise in pseudowhorls **(Fig. 5.1A).** They are dimorphic with long shoots (branches of unlimited growth) and dwarf shoots or brachyblast (branches of limited growth). Leaves are also dimorphic – foliage leaves or needles and scale leaves.
- *Pinus* has a well developed tap root. It remains short and grows on hard ground or rocks. The lateral roots spread over a large area. Young roots are infested with a fungus to form mycorrhizae.

Morphology

Branches

- Dwarf shoots or foliar spur or shoots of determinate growth are borne on long shoots in close spirals from axils of scaly bracts and bear fascicles of leaves (needles).
- The long shoots or shoots of indeterminate growth appear on the main stem as lateral buds in the axils of scale leaves and spread horizontally **(Fig. 5.1B).** They shorten gradually towards the tip giving tree a pyramidal appearance. Each of these shoots terminates in an apical bud, which is enclosed by a number of bud scales closely surrounded by a thick mat of hair. Buds are ovoid to cylindric with pointed apex, are usually resinous.

Leaves

- Scale leaves, similar to bud scales, are small, brown and non-photosynthetic, and arranged spirally like the juvenile leaves. The scale leaves are firmer at the base and persist for life of fascicle or shed after first season.
- The foliage leaves or the needles are borne on dwarf shoots axillary to scales in clusters or fascicles of one to eight needles. They are initially bound together by a basal sheath that may fall off (with needles) or may persist. The number of leaves per fascicle, the length of needles, the number of sides of the needles (only *Pinus monophylla* has a round needle), the distribution of stomata (waxy white specks on the leaf surface), and the colour and stiffness of the needles can all be useful

characters for identification. Foliage leaves or needles, two to five in number or it may be minimum one and maximum six needles per fascicle. They may persist for two to twelve or more years, are sessile, sheathed at base by twelve to fifteen overlapping scale leaves.

- Since plant continues to grow for several years, it bears five types of leaves in succession during its life span: Cotyledonary leaves, Juvenile or primary leaves, Prophylls, Cataphylls and Acicular leaves.

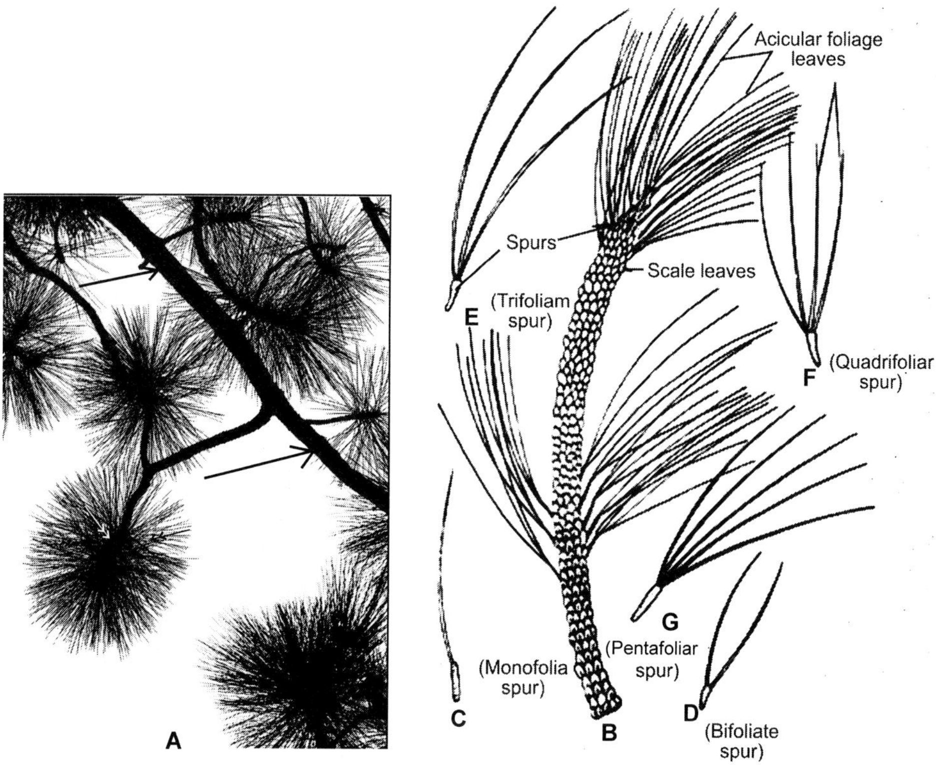

Figure 5.1 A-G: *Pinus* Sporophyte.

A - Portion of the canopy showing long shoots (arrows) and dwarf shoots (double arrows). B - Long shoot bearing dwarf shoots or spurs. The spurs bear a specific number of foliage leaves (C-G).

- Cotyledonary leaves are the first to be produced when the plant emerges from the seeds. There may be from four to twenty four of them (24 cotyledons, from *Pinus maximartinezii*, is the largest number known in any plant).
- Primary or Juvenile leaves are single, alternate (usually helically arranged), broad with quadrangular outline that are usually produced only for the first year of growth but that may be produced for many years in some species (such as *Pinus quadrifolia*). When the pine is two to three

years old, all the primary leaves are shed off.

- The other three types of leaves are present on dwarf shoot: (a) acicular or foliage leaves (needles) – two to five per fascicle, terete or two to three angled and rounded on abaxial surface, sessile, (b) scale leaves or prophylls, non-chlorophyllous primary leaves, typically small, subulate or lanceolate, with ciliate margins and are protective in nature. Two opposite scale leaves form a sheath at the base of foliar spur and (c) cataphylls – five to thirteen overlapping, spirally arranged scaly cataphylls are present next to scale leaves and arise in 2/5 phyllotaxy.
- The number of needles per spur is constant for a species **(Fig. 5.1B-G)** and is a taxonomic character such as: *P. monophylla* (one needle), *P. sylvestris* (two needles), *P. roxburghii* (three needles) and *P. wallichiana* (five needles).
- During the reproductive stages, male cones arise laterally on long shoots or terminally on dwarf shoots. Female cones arise on separate shoots, appear woody at maturity and may be small to very large in size depending on species.
- Seeds are slightly flattened and usually winged (with one wing).
- Seedlings with varying number of cotyledons, from two to twenty four, arise when seed germinates.

Anatomy

Stem

- The outline is not circular due to the leaves that are present on stem, make it undulated and scaly **(Fig. 5.2A).**
- The cuticularized epidermis is followed by a hypodermis and broad parenchymatous cortex.
- The endodermis and pericycle demarcate a ring of vascular bundles.
- The vascular bundles are conjoint, collateral and open.
- They form a ring around the pith. In each vascular bundle, the xylem is towards the inner side and phloem towards the outer side.
- Medullary rays present between the vascular bundles are narrow.
- Resin canals are distributed in the cortex and xylem or wood **(Plate V.1).**

Secondary Growth

- Occurs as in dicot stems by means of interfasicular and intrafasicular cambium.

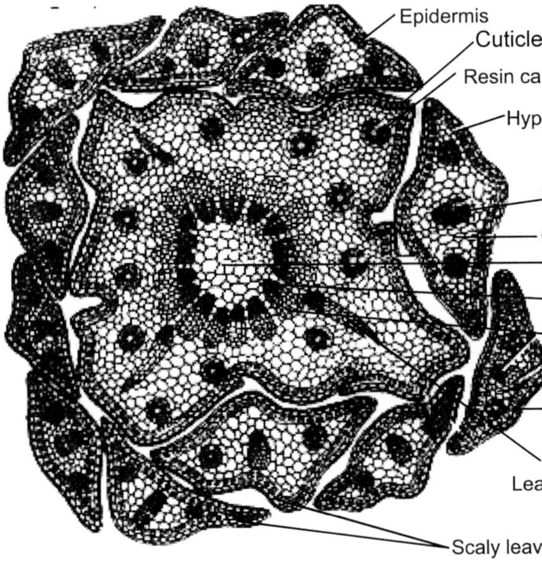

A

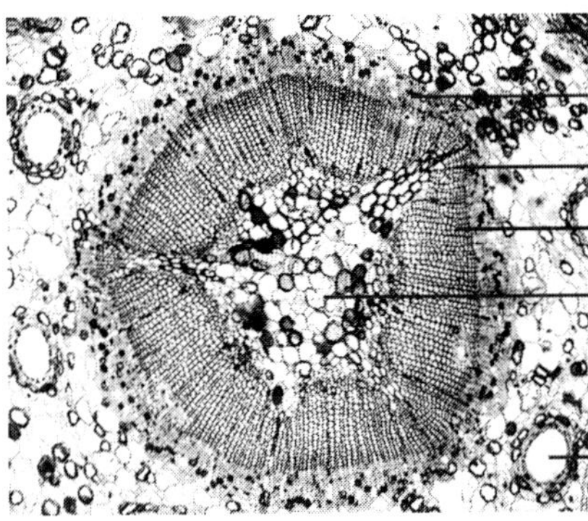

B

Figure 5.2 A&B: *Pinus* T.S. Stem.

A - Shows primary structure of the long shoot. The small pith and small cortex seen along with a ring of vascular bundles that occupy more space, is a character of a pycnoxylic wood. Also seen are scale leaves on the shoot that are cut in transverse section. B - Resin canals in the cortex are abundant. The resin canals are lined with a distinct epithelium which secretes resin into the canal.

- This cambium cuts off secondary phloem towards the inside and secondary xylem towards outside **(Fig. 5.2B)**.
- There are no vessels but xylem has tracheids with bordered pits on radial walls.

- The secondary growth is more pronounced in spring and not so in summer/autumn, and the size of the tracheids differs, being larger in spring wood.
- As a result of this alternate spring-summer secondary tissue formation, growth rings are formed. One spring-summer growth is equal to an annual ring.
- As the plant ages, the wood shows a lighter zone of sap wood and a darker zone of heart wood.
- The primary medullary rays are replaced by secondary medullary rays with a complex structure and with greater significance in gymnosperms than angiosperms.
- They run through secondary xylem and secondary phloem; the former is xylem ray and latter is phloem ray.
- The xylem ray consists of a single row of starch cells bounded on either side by radially elongated thick walled cells with simple pits. These are the tracheidal cells which help in radial diffusion of water. The phloem ray is composed of central rectangular starch cells and peripheral albuminous cells. These cells are rich in proteins.
- The rays are of two types: uniseriate and multiseriate or fusiform. These can be examined in T.S., T.L.S. and R.L.S. **(Fig. 5.3)**. Transverse section (when cut at right angle to stem axis), tangential (when cut parallel to

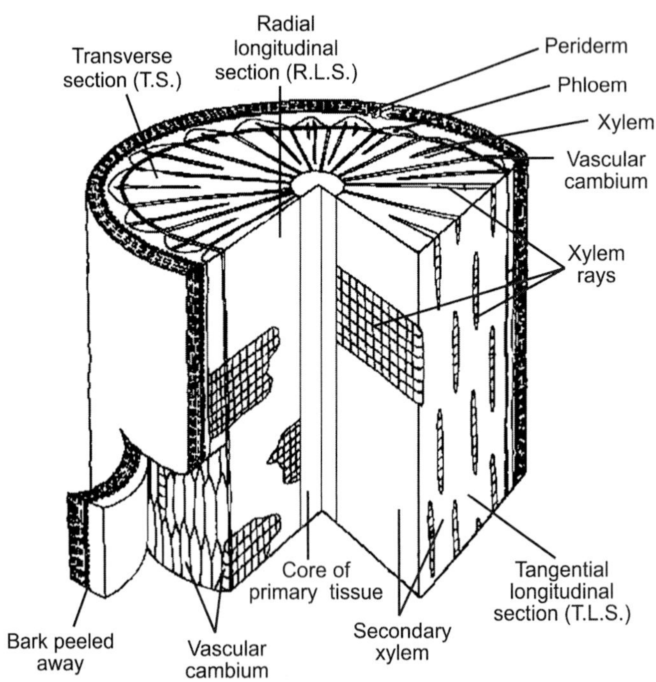

Figure 5.3: *Pinus* various planes of sectioning of wood (http//www.ups.gov).

the stem axis away from the centre of axis), and radial (when cut parallel to stem axis and exactly through the centre of axis) respectively. Each section shows a view of different angle of tissues and cells, and represent particular features that are helpful in the identification of the wood.

- Secondary growth in the cortical region is due to phellogen that arises in the cortex resulting in a thick scaly bark but some species have thin, flaking bark.
- Pith remains small.

Wood

- In old wood, the outer lighter zone, sapwood, and the inner darker region, the heartwood are easily demarcated.
- With age, parenchyma cells begin to die and sapwood gets gradually converted into heartwood which becomes resinous and rich in phenolic compounds. This makes wood more compact and durable.
- Distinct concentric circles of annual or growth rings are formed by alternate formation of vigorously produced large thin walled cells in the early or the spring wood and thick cells of the late or the summer wood. So the age of a conifer tree can be determined by counting the number of annual rings **(Fig. 5.4)**.
- *Pinus* has pycnoxylic wood **(Fig. 5.2 A&B)**, very much simpler as compared to wood of dicotyledons.
- Wood consists mainly of tracheids and vessels are absent.
- Another characteristic feature is the absence of fibers, however, old wood may have fiber tracheids. Wood parenchyma is absent.
- Ducts leading to resin canals are present in the wood region. The resin ducts are lined by epithelial cells. They are associated with turpentine and resin secretions.
- All the above described features make pine commercially important as a timber yielding plant.

**Pines are divided into two groups, five needle pines (soft pines) and two or three needle pines (hard pines). One of the common occurrences in pine is compressed wood. It is formed along the under surface of branches and on trunk of leaning trees.*

Twenty-two, thirty-one, and twenty-eight compounds have been identified in the essential oils of wood, bark, and needles, respectively. The major chemical constituents of wood's essential oil were caryophyllene (16.75%), thunbergol (16.29%), 3-carene (14.95%), cembrene (12.08%), α-thujene (10.81%), and terpinolen (7.17%). In bark, α-pinene (31.29%) and 3-carene (28.05%), and in needles, α-pinene (39%) and 3-carene (33.37%) were found. Almost all of the essential oils were active against pathogen bacteria.

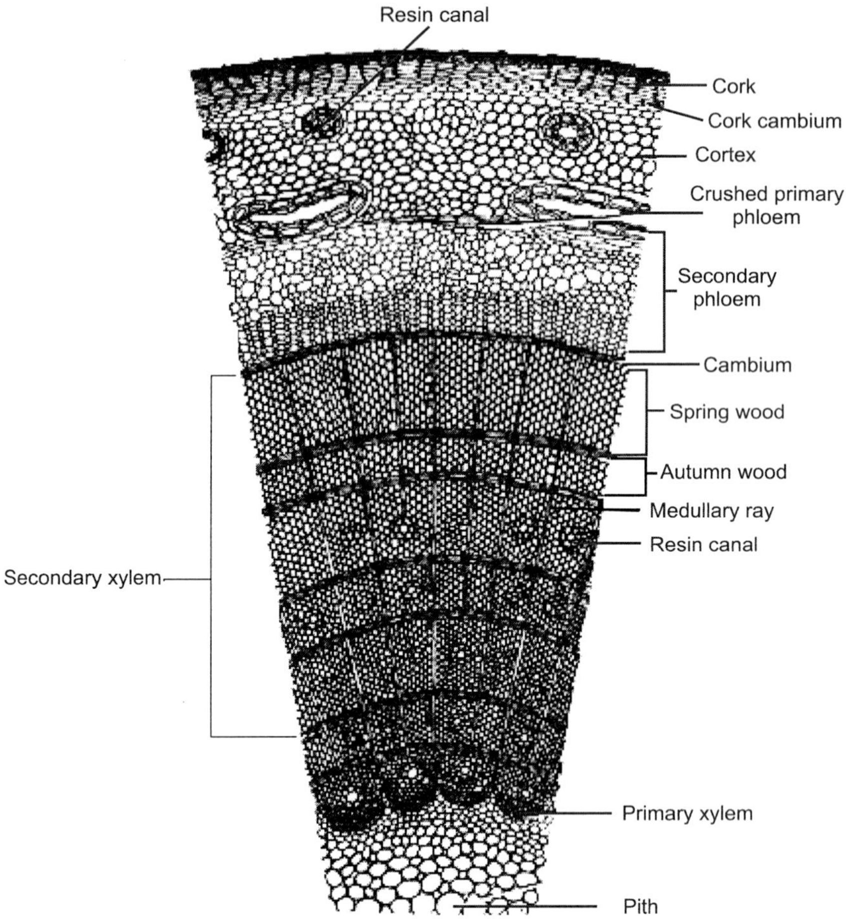

Figure 5.4: *Pinus* T.S. Old Stem, cellular details show secondary tissues with distinct growth rings formed by spring and summer wood. Both secondary wood and the secondary cortex show resin canals. Note the distinct medullary rays in the wood and the resin canals lined by epithelium are distributed in the cortex. Source: http://istudy.pk/wp-content/uploads/2016/11/TS-of-stem-1.jpg.

T.L.S. Stem

Shows tracheids and medullary rays.

- Tracheids are composed of rectangular cells and have bordered pits.
- Each bordered pit is enclosed by a pit chamber bounded by a pit membrane and contains a centrally located swollen torus.
- The rays may be uni - or multiseriate (**Fig.5.5A**).

- The height of uniseriate ray varies from one to twelve cells.
- The multiseriate ray is always more than one cell wide and several cells in height. It is associated with centrally situated resin canal.
- Pith is not visible in this plane.

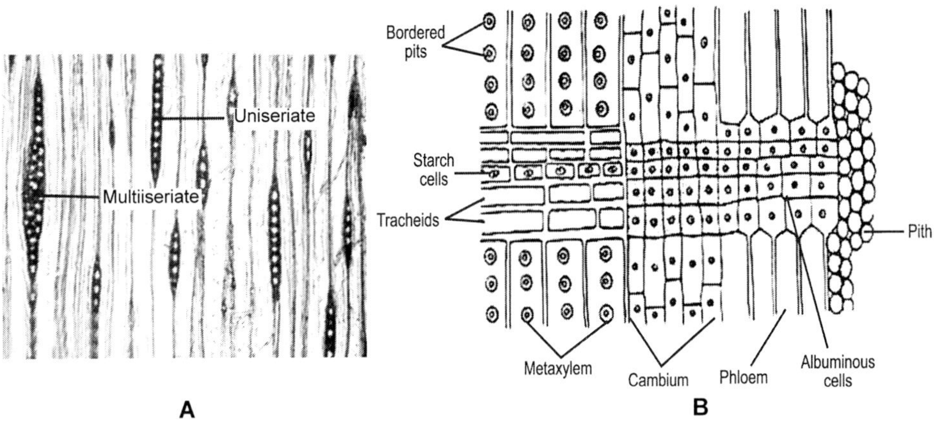

Figure 5.5 A&B: *Pinus.* Stem.

A - T.L.S. Showing uniseriate and multiseriate rays. Also seen are bordered pits. The height of uniseriate ray varies between 6-8 cells. B - R.L.S. with medullary rays that run perpendicular to xylem and phloem.

R.L.S. Stem

Since the section passes through the centre it shows xylem, and phloem and perpendicular to these tissues are seen medullary rays – xylem ray and phloem ray.

- Bordered pits of tracheids are seen surrounded by Bars of Sanio in surface view **(Fig. 5.5B)**
- Uniseriate medullary rays run horizontally and have tracheidal cells on either side which are associated with starch cells. Thin walled ray parenchyma is also present. Xylem is separated from phloem with cambium.
- Phloem consists of sieve tubes, sieve plates and phloem parenchyma and medullary rays. Latter are associated with albuminous cells, which are frequently equated to companion cells of the angiospermic phloem **(Fig. 5.5 B)**.
- Pith is visible as the section is through radius of stem **(Plate V.2)**.

Root

- The outermost tissue is epiblema or a piliferous dark layer, followed by a large parenchymatous cortex.
- In young roots ectomycorrhizal mycelia are seen.
- The stellar region is demarcated by endodermis with about three to four layers of pericycle inner to it.
- Vascular bundles are radially arranged and diarch to tetrarch with exarch protoxylem **(Fig. 5.6)**.

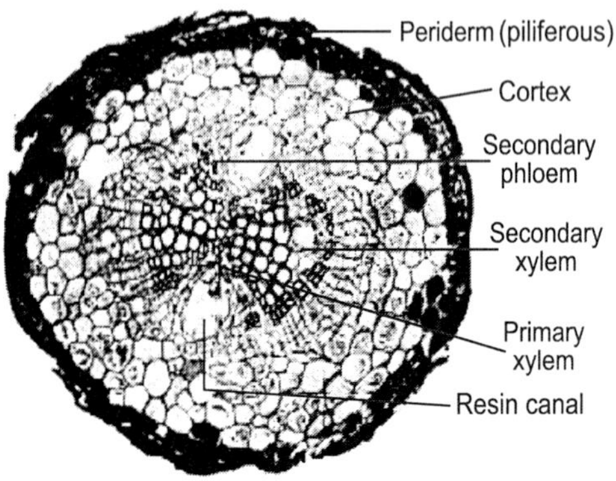

Figure 5.6: *Pinus* T.S. Root cellular details show a diarch condition and outer surface appears dark and piliferous.

- Protoxylem is bifurcated (Y-shaped) towards the periphery, and in between each bifurcation a resin canal is present.
- Phloem is present alternate to the protoxylem.
- Pith is poorly developed or absent **(Plate V.3)**.
- Secondary growth sets in when the primary tissue is still in the process of differentiation.
- Cambium develops beneath the phloem and cuts secondary xylem towards the pith and secondary phloem towards the cortex.
- Cork cambium also develops in outer region of pericycle and cuts off cork cells externally and parenchyma cells internally.

Leaf

The leaf number varies on the shoot of limited growth **(Figs. 5.1B & 5.7A)**.

- Leaf is covered by thick walled epidermis which is interrupted by sunken stomata leading to a substomatal cavity.
- Outer to epidermis is a thick layer of cuticle.
- Two or three layered sclerenchymatous hypodermis is present below the epidermis.

- Mesophyll of the leaf is parenchymatous. It is not differentiated into palisade and spongy parenchyma. To increase surface area for photosynthesis, parenchyma shows folded membrane resembling 'pegs' and is called 'flanged or armed' parenchyma **(Plate V.4).**
- Resin canals are present below the hypodermis.
- Two conjoint, collateral and closed vascular bundles are present in the centre.
- Endodermis is present outside the pericycle.
- Pericycle is multilayered and is much differentiated. It consists of mainly parenchymatous cells but is sclerenchymatous forming T-shaped girder, which separates two vascular bundles. It is also represented by transfusion tissue present adjacent to xylem and consists of tracheidal cells while pericycle cells adjacent to the phloem form albuminous cells. These specialized cells form the transfusion tissue.

Thick cuticle, sunken stomata, well developed sclerenchyma, transfusion tissue and flanged mesophyll are characters of xerophytic leaf* *(Fig. 5.7B,C).***

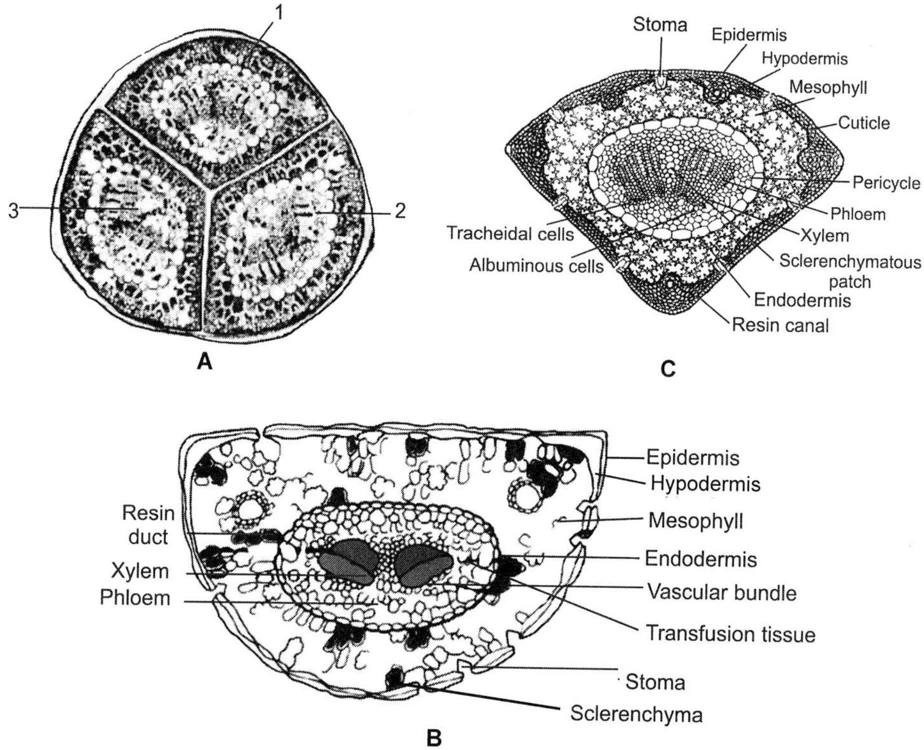

Figure 5.7 A-C: *Pinus* T.S. Needle.

A - Trifoliar spur shows three needles (1, 2, 3). B - Diagrammatic sketch. C - One needle enlarged to show sunken stomata, sclerenchymatous hypodermis and transfusion tissue adjacent to two veins. Note the flanged mesophyll.

Reproduction

There is no vegetative reproduction. Plant reproduces sexually and forms monosporous cones. Both male and female cones are formed on the same plant but different branches **(Fig. 5.8).**

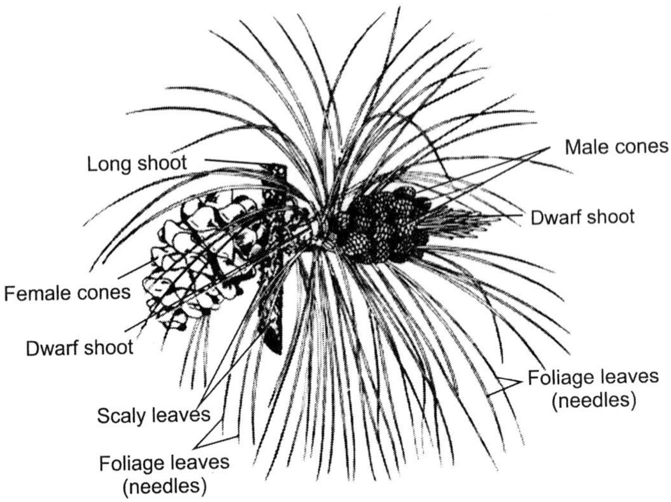

Figure 5.8: *Pinus* Reproductive Structures.
Both male and female cones are borne on the same plant but different branches.

Male Branch

The male cones are produced in clusters in the axils of scale leaves at the base of the youngest shoot of the unlimited growth of the current year **(Fig. 5.9A).** The main shoot continues to grow forming more dwarf shoots in the axils of the upper scales **(Fig. 5.9B).** Thus, male cones replace the dwarf shoots. They are shed after pollination.

Male Cone

- The male cones are produced in spring.
- They are more in number, smaller than female cones and present in clusters **(Fig. 5.9C)** in the axils of scale leaves of the long shoots. The main shoot continues to grow.
- Young cones may have a distinctive colour, but normally identification requires mature cones **(Plate V.5).**
- Each male cone has a central axis that bears spirally arranged microsporophylls.

Figure 5.9 A-C: ***Pinus*****. Male reproductive structures.**

A - Male branch to show cluster of young male cones in axil of scale leaves. Also seen is a pendulous old cone. B - The growth of the shoot resumes (arrows) after the cones are formed. C - Cluster of male cones is enlarged.

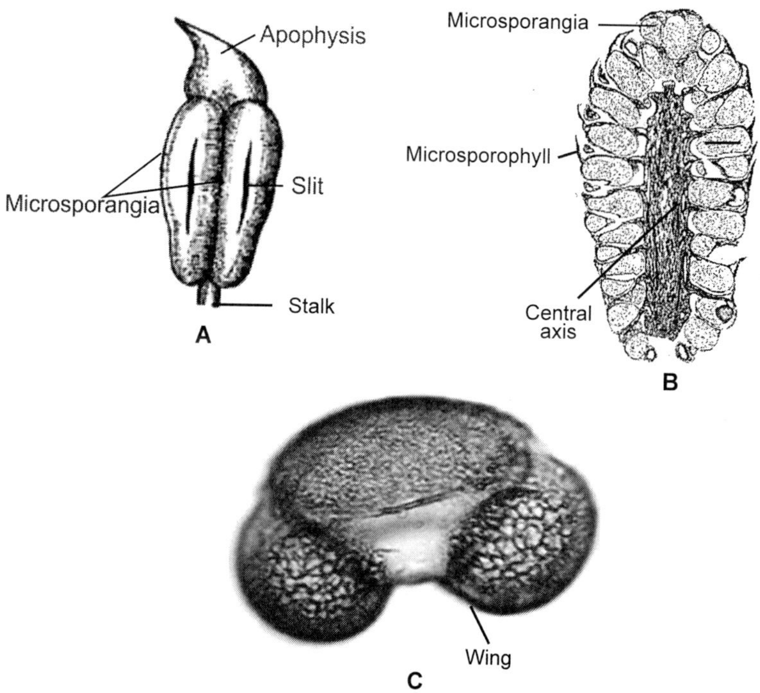

Figure 5.10 A-C: ***Pinus*****. Male reproductive structures.**

A - Microsporophyll is seen with two microsporangia. The sterile tip, apophysis is distinctly visible. B - L.S. of the cone to show central axis and microsporophylls. The microsporangium with microspores are seen to be borne by the microsporophyll which has a pointed sterile tip. C - Shows a winged pollen grain.

- Each microsporophyll has sac-like microsporangia (pollen sacs) on the ventral side. The tip of the sporophyll is triangular with pointed end curved up and gives protection to the microsporangia **(Fig. 5.10A; Plate V. 6&7).** Several microsporophylls bearing microsporangia are arranged around the central axis **(Fig. 5.10B).**
- Each microsporangium has a distinct wall and produces a large number of winged microspores **(Fig. 5.10C)**.

Microsporangium

- One or more hypodermal cells of the microsporophyll act as sporangial initials and divide to form outer wall initials and the inner archesporial initials.
- The wall initials give rise to wall of the microsporangium.
- The archesporial initials divide and form archesporium that later develops into microspore mother cells.
- The peripheral cells of the archesporium form the tapetum.
- Tapetum is secretory type and provides nourishment to the developing microspore mother cells.
- The microspore mother cell divides by meiosis to form four microspores or pollen grains.
- Microspore at maturity has an outer wall exine and inner intine. The exine forms wings or sacci that help in dispersal **(Fig. 5.10B,C).**

Female Branch

Ovulate cones are borne in the axils of scale leaves at the apices of the young upper branches of unlimited growth (**Fig. 5.11).** Each female cone replaces a long shoot.

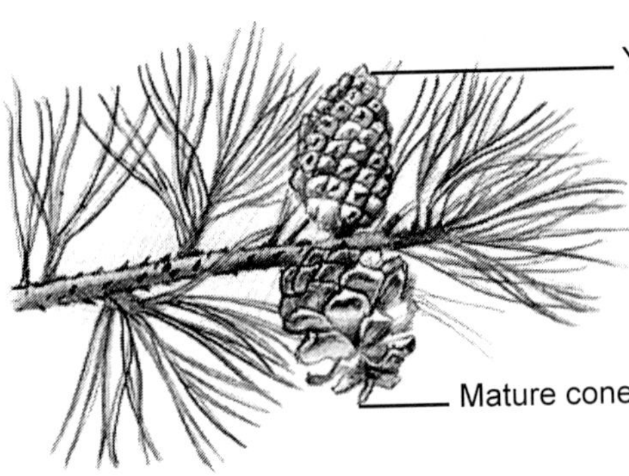

Figure 5.11: ***Pinus.*** **Female reproductive structures. Young and old cones on long shoot.**

Female Cone

- Female cones start appearing in winters and become receptive by following spring.
- Ovulate cones are solitary to a few in number.
- They mature in about two or three years, and are shed early or may stay on the branch for some time. They are pendent to erect, and the shape at maturity is conic or cylindric **(Plate V.8)**.
- Though some cones shed seeds soon after maturity others may require a stimulus as in serotinous cases (not opening upon maturity but much later, usually in response to fire).
- Each young female cone has a central axis bearing spirally arranged pair of scales known as bract-scale complex **(Fig.5.12A, Plate V.9)**.
- The bract-scale complex consists of two types of scales: (a) bract scale which is thin membranous and is directly attached to the central axis and (b) ovuliferous scale which is persistent, woody and present on the ventral surface of each bract scale **(Fig. 5.12B&C)**.
- The broader end of the ovuliferous scale has projection called the umbo.
- At the base of upper surface of each ovuliferous scale are present two sessile and naked ovules **(Plate V.10 A&B)**.

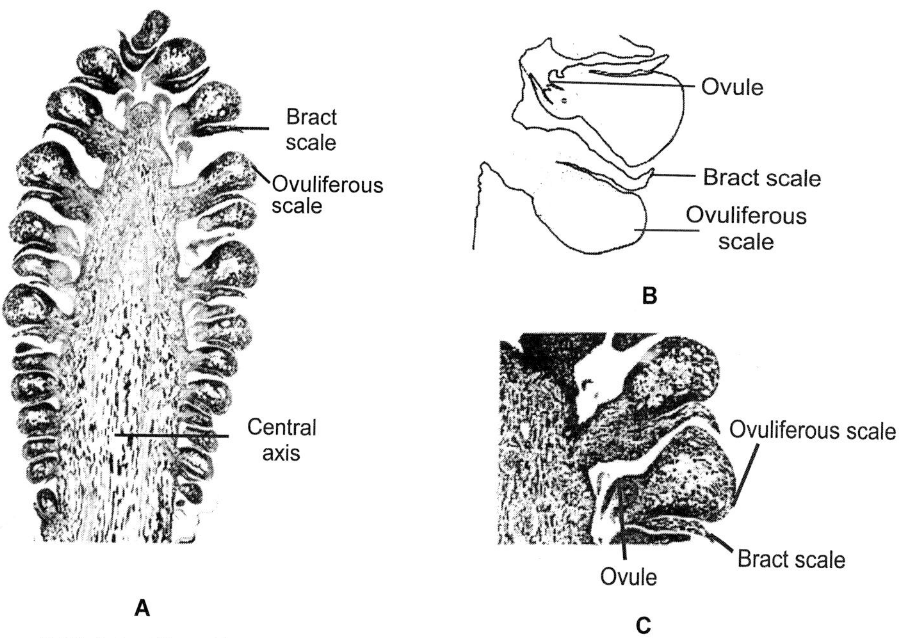

Figure 5.12 A-C: *Pinus* Female Cone.

A - L.S. to show central axis with scale-complex. B - Outline representation of scale complex. C - Scale-complex is enlarged to show a bract scale and an ovuliferous scale which bears an ovule at the base. The tip of the scale is extended.

Cones take two or three years to mature **(Fig. 5.13A-C; Table 5.1).** The time requirement varies between species. Cone scales may remain closed and sealed by resin long after cone maturity; such cones are adapted to open when the resin melts by fire and are called serotinous. Normally, cones open at maturity and release seeds **(Fig. 5.13C)**. The seed may be wind-dispersed, in which case it is normally small and light with a wing longer than the seed (**Fig. 5.13D**); in some, the seed is normally larger and heavier, and the wing may be reduced to inconspicuity.

★Female cone matures in about one to three years and their comparison is tabulated.

Figure 5.13 A-D: *Pinus*. Female cones at various stages of development.

A - First year cone is soft and fleshy. B - Pollinated closed cone of second year. C - Mature third year cone where the scales have opened to release the seeds. D - Shows winged seeds.

Table 5.1: Comparison of female cone at various stages of development.

	1st Year	2nd Year	3rd Year
Shape	Oval	Elongated and larger than the first year cone.	Elongated or roughly rounded.
Length	1 to 4 cm	5 to 15 cm	Not much difference but is more open.
Colour	Green to reddish-green	Reddish brown	Brown
Megasporophyll arrangement	Megasporophylls are arranged very close to each other, and so the cone is a compact structure.	Megasporophylls are compactly arranged, but not so compact as in 1st year cone.	Megasporophylls are very much loosely arranged.

Megasporangium

- The ovule (megasporangium) is anatropous, has a mass of nucellar tissue which is surrounded by a single integument (unitegmic) **(Plate V.10B).**
- The micropylar end of the ovule is directed towards the central axis.

- A pollen chamber is present in which the pollen germinates.
- A single megaspore mother cell is differentiated in the nucellus near the micropylar end.
- This megaspore mother cell undergoes meiosis to form four haploid megaspores.
- The lower most megaspore remains functional while others degenerate.
- Functional megaspore increases in size. It occupies the major part of the nucellus and is the progenitor of the female gametophyte. Overwintering occurs at this stage.

Gametophytes

Male Gametophyte: Pollen Grain

- The haploid microspore nucleus (**Fig. 5.14A**) divides into a small lens-shaped prothallial cell and a large central cell at distal end.
- Further, central cell gives rise to second prothallial cell and antheridial cell initial.
- During further development, the antheridial cell initial cuts off antheridial cell and tube cell **(Fig. 5.14B).**
- The winged pollen grain at the time of dispersal/pollination is four-celled with two small lens-shaped prothallial cells towards proximal end, antheridial cell and a tube cell (**Fig. 5.14C**).
- Both the prothallial cells are ephemeral.

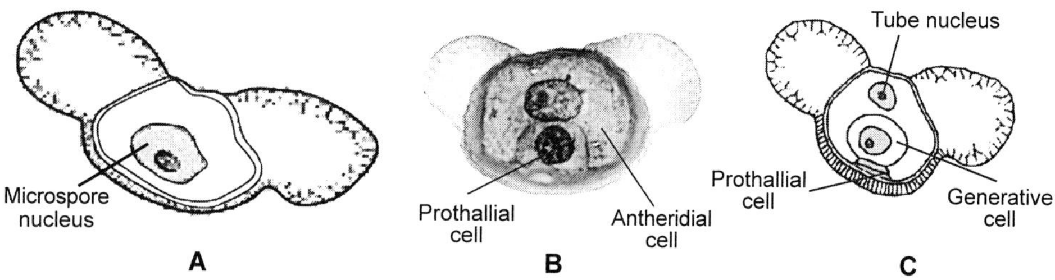

Figure 5.14 A-C: *Pinus.* The male gametophyte development. From a uninucleate stage, the pollen grain passes bicelled stage and finally it is shed at four celled stage.

Female Gametophyte

After overwintering, development in the following spring (one year after pollination) is rapid. More mitotic divisions in the megagametophyte result in a free nuclear state of about 1024 or 2048 haploid nuclei before wall formation begins.

- Functional megaspore increases in size and its nucleus undergoes a series of free nuclear division.

- A vacuole develops in the centre with about 2000 nuclei arranged towards periphery.
- After a certain number of nuclei are formed, cell walls are laid down which ultimately result in obliteration of the vacuole.
- Wall formation starts from periphery and proceeds towards the centre. The first walls are laid perpendicular to the megaspore membrane and extend to the middle of the central vacuole. These cells are like long tubes that are open towards the inside and are called 'alveoli'.
- As a result of laying down of walls towards the center, the entire gametophyte also known as endosperm becomes cellular.
- It is a pre-fertilization tissue.

Pollination

In the months of May and June, pollen grains are transferred by wind. Male cones are borne in clusters and there is a large number of microsporangia in each cone, bearing many microspores. In day time when pollination occurs in pine forests the yellow dusty pollen grains fill up the air. This is called sulphur shower **(Plate V.11)**.

Post Pollination Events

Male Gametophyte

The four celled pollen rests on the nucellus before it begins to germinate. A pollination drop is secreted by the ovule and once the drop dries up pollen grains are taken in where they land in the pollen chamber.

- Tube nucleus controls growth of the pollen tube.
- Further development of pollen grain takes place at the surface of the nucellus.
- Antheridial cell also cuts off spermatogenous cell (or the generative cell) and stalk cell.
- Exine of the pollen grain ruptures.
- Intine forms the pollen tube which grows through the nucellus. But its further activity stops till spring.
- The spermatogenous cell gives rise to two male gametes just prior to fertilization **(Fig. 5.15A-F)**.

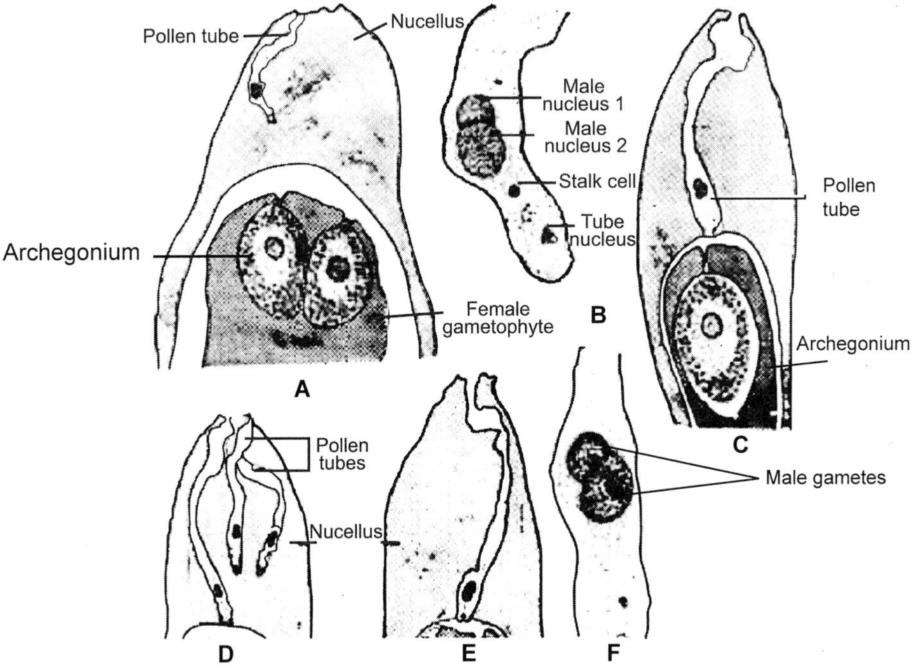

Figure 5.15 A-F: *Pinus*. Female gametophyte, post-pollination changes.

A - Germinated pollen tube travels through the nucellus to make contact with the egg cell. B - The final development and formation of male gamete is seen where the spermatogenous cell has divided forming two sperms. C - Pollen tube has contacted the archegonia. D - Several pollen tubes arise, a factor responsible for polyembryony. E-A closer view to show that two male gametes are positioned near the tip of pollen tube. F - Shows two sperm nuclei.

Female Gametophyte

With pollination stimulus, archegonial development is initiated in the gametophyte.

- After pollination ovuliferous scale enlarges in size and bends downwards so that the cone hangs vertically. Outer ends of the ovuliferous scales increase in size until they meet each other to close the gaps between them. The cone is covered with a lot of resinous secretions. It may be noted that ovuliferous scale and bract-scale which make a bract-scale complex, has been referred to as bract-seed-scale complex when the ovules borne by the ovuliferous scale have attained maturity and grow into seeds.
- Few cells (two to four) at the micropylar end of the endosperm or female gametophyte become large and prominently accumulate proteins and RNA, functioning as archegonia initials. Each archegonial initial soon divides into a large central cell and a small primary neck initial cell. Archegonia develop with a large venter and a short neck.
- The central cell enlarges very rapidly because of which numerous vacuoles are formed and structure is referred to as 'foam stage'.

- Later central cell forms an ephemeral ventral canal cell and a large egg cell.
- The neck is without any neck canal cells.
- The gametophytic cells near the apical end of individual archegonium grow vigorously resulting in sinking of the latter **(Plate V.12).** Each archegonium acquires its own archegonial chamber. Cells surrounding the archegonium form a special covering layer, the jacket.
- Archegonium wall shows pits with further development **(Fig. 5.16 A&B).** Through these pits the egg cell haustoria penetrate and reach the jacket cells for nutrition (https://dc.lib.unc.edu/cgi-bin/showfile).
- The developing gametophyte enlarges in size, crushing all nucellar tissue except a small amount near the micropylar end.
- The egg nucleus enlarges and descends to the centre of the egg cell. Although a particular ovule may produce only one archegonium, the number is usually higher and multiple archegonia are seen. The maximum number varies from two to seven. This is an important feature of mating systems of pines and explains high levels of variations in the genus.

In *Pinus radiata* post pollination changes have been studied in detail. Immediately prior to and during pollination cones emerge from the bud scales and their axes elongate. As a result, the scale complexes are pushed apart and become open exposing the ovules. At the same time ovuliferous scales swell and become darker in colour. Cones are regarded as fully receptive when their scale complexes are horizontal and the channels between two neighbouring ones are wide. The ovuliferous scales in the upper part of the cone bear two ovules at their bases and the ovules project fine micropylar end into the channels between the scales. The ovuliferous scales of the lower part of the cone are sterile bearing either rudimentary ovules or none at all. Wind carries many of the pollen grains into the channels of the ovules where they are intercepted by the micropylar arms of the ovule. The nucellus secretes a pollination droplet **(Plate V.13)** between the arms of the micropyle, and pollen lodges on the sticky surface of the droplet, which is fairly high in glucose, fructose and sucrose (McWilliam, 1958). The pollen gets attached to the ovules either because of sticky pollination drop (Doyle and O'Leary, 1935) or through a bioelectric effect (McWilliam, 1959). The droplet is withdrawn, apparently reabsorbed, drawing any trapped pollen into a depression on the surface of the nucellus called the nucellar or pollen chamber. The middle cell layer in the micropylar neck enlarges, closing the neck. If pollen is plentiful, to or four pollen grains may be lodged in each pollen chamber, a very significant factor in the genetics of the pine. The pollen grains germinate and begin to penetrate the nucellar tissue **(Fig. 5.15A-F).** Pollen is then carried inside the ovule for germination by a tube leading from space between the micropylar arms to the cavity in the nucellus cap of the ovule. The means of transference is assisted by pollination drop (Doyle and O'Leary, 1935; McWilliam, 1958; Sarvas, 1962). Cone closure after receptivity is due to the continued swelling of the ovuliferous scales which eventually meet and seal the cone. After germination, the generative nucleus divides to form the stalk and body cells or may directly give rise to sperms (different views). Further development proceeds slowly over the summer.

While archegonia are developing, the pollen tubes grow rapidly in the direction of megagametophyte and the body cell divides to form two unequal, unflagellated sperms. The tip of the pollen tube forces through the neck of the archegonium and discharges its contents into the receptive vacuole in the egg.

During embryogenesis, megagametophyte continues to grow and provides the nutritive tissue often called endosperm for the developing seedling.

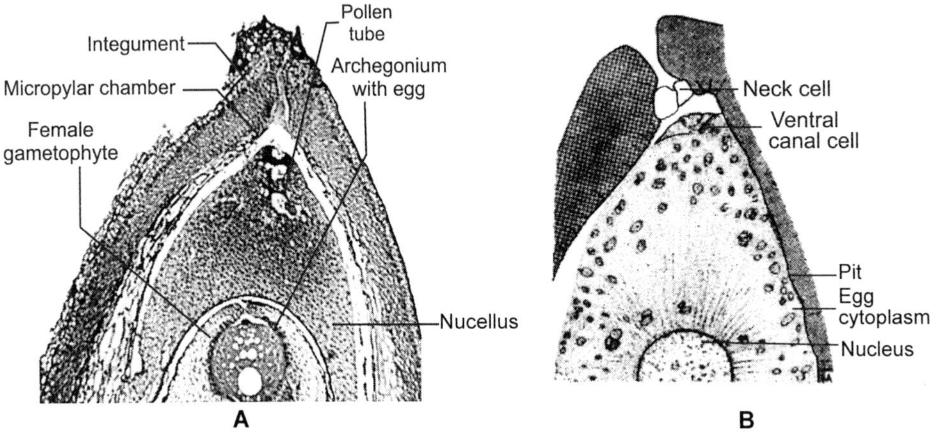

Figure 5.16 A&B: *Pinus* L.S. Ovule with post pollination changes.

A - L.S. Micropylar end shows archegonium and pollen chamber. B - The archegonium enlarged to show pits in the jacket. The neck is inconspicuous when compared to venter.

Fertilization

The cone remains closed and the male and female gametes complete their development before fertilization.

The pollen tube formed in the pollen chamber of ovule becomes active and after piercing through nucellus reaches the neck of archegonium. Tip of pollen tube now contains two male gametes and the tube and stalk nuclei. Pollen tube comes in contact with archegonium and the tip ruptures discharging the contents into the egg. One of the male gametes approaches the egg nucleus and unites to form zygote. Second male gamete along with the stalk and tube nuclei disintegrates.

Fertilization takes place in June and July. More than one pollen tube may develop in association with a single ovule and more than one archegonium may be fertilized.

Since several pollen grains are trapped in the pollen chamber and multiple archegonia are formed, it results in multiple embryos, a condition called polyembryony. Competition occurs between embryos and usually only one survives to maturity.

Embryogeny

The development of the embryo in *Pinus* can be divided into three phases (Singh, 1978); (1) Proembryogeny – before suspensor elongation. (2) Early embryogeny – all stages during and after the elongation of suspensor but before the establishment of root meristem. (3) Late embryogeny – establishment of root and shoot meristem and further development of the embryo.

Proembryogeny starts with a free nuclear stage. The zygote undergoes free nuclear divisions. After cell wall formation, four tiers are formed of which the lowest tier (E) will form the embryonal mass and the second lowest (E1) will form the embryonal suspensor (**Fig. 5.17A-D)**. The phase of proembryogeny ends just as the E1 tier starts to elongate.

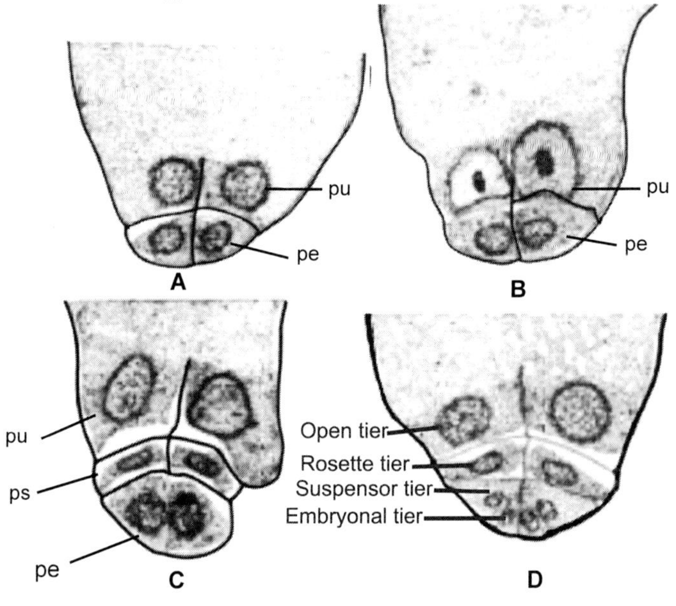

Figure 5.17 A-D: *Pinus*. Early stages in embryogeny.

After fertilization, the development of the zygote leads to formation of a proembryo which shows distinct pro embryonal tiers - embryonal, suspensor, rosette and open tiers. Abbreviations: pe = primary embryonal tier, ps = primary suspensor tier and pu = primary upper tier.

Early embryogeny, the E1 tier continues to elongate and the E2 tier gets organized which later splits or cleaves longitudinally into four disjoined tiers **(Fig. 5.18A-C)**, giving rise to so-called cleavage polyembryony (Grifford and Foster, 1989). Each of the four disjoined E tiers gives rise to an embryo that begins the development by apical cell growth (Buchholz, 1920). The embryos start to compete for nutrients and space to grow and survive. However, it is not only a competition among the four embryos within the same archegonium, but also with those of the other archegonia within the same seed. Usually, only one embryo, the dominant embryo, wins the competition and reaches maturity while the others, the subordinate embryos, are suppressed, arrested and subsequently are eliminated. During early embryogeny, the most basally situated cells in the embryonal mass undergo asymmetric division to form embryonal tube cells (Et). These are subsequently added to the suspensor system as new tube cells are formed. Thereby, several files of secondary suspensor cells are continuously formed (E2, E3...*etc.*), leading to a rapid growth of the suspensor system (Chowdhury, 1962; Filonova *et al.,* 2002). The suspensor cells do not divide but they elongate and become highly vacuolated (Singh, 1978; Smertenko *et al.,* 2003). The rapid growth of the suspensor system forces the apical tier out of the archegonium and into the megagametophyte (Chowdhury, 1962; Skinner, 1992) from where they draw their nourishment. Each tube carries a mass of embryo at its tip. Accessory embryos are believed to arise from rosette tier. This is type of polyembryony.

Late embryogeny is the phase of histogenesis and organogenesis during which Root Apical Meristem (RAM) and Shoot Apical Meristem (SAM) are organized. Finally by seed maturity, SAM is surrounded by a whorl of cotyledons and the RAM is

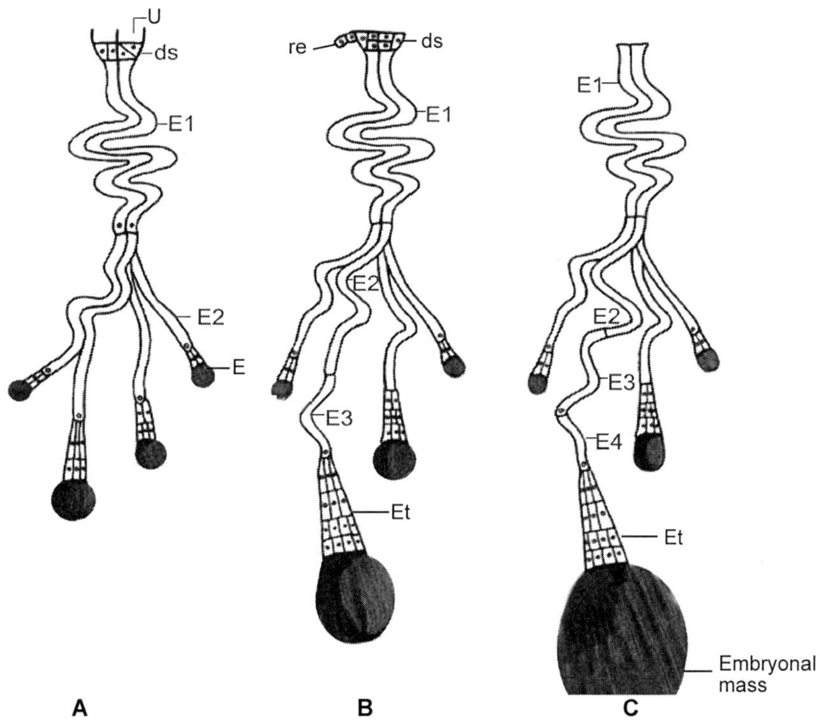

Figure 5.18 A-C: *Pinus.* Stages in late embryogeny.

Cells of various tiers have elongated. The primary (E1) and secondary suspensor cells (E2) elongate and push developing embryonal mass (E) deep into the gametophyte. Note the developing rosette embroys from ds (dysfunctional suspensor) tier. The embryonal tubes carry embryonal tier which undergoes divisions to organize embryos. Abbreviations u = upper tier and open tier, ds = dysfunctional suspensor tier or rosette tier, E1 - embryonal tier or primary suspensor tier and E2 = secondary suspensor tier; E = embryonal tier, re = rosette embryos, et= embryonal tubes

covered by a root cap (Raven *et al.,* 1999). *Pinus* species can show meristematic cells in the dS tier (distal tier). The cells in the dS tier do not elongate as the secondary suspensor cells (Dogra, 1967; Singh, 1978), however, they can start to divide to form lobes, resulting in the formation of so-called 'rosette embryos' (Chowdhury, 1962). These cell masses never form embryos, but are instead eventually aborted. The important stages in pollination **(Fig 5.19A)** and embryogeny **(Fig 5.19B&C)** are depicted for clarity of conept.

Seed

- Seed is a winged structure **(Plate V.14).**
- The seed coat shows differentiation into testa and tegmen.

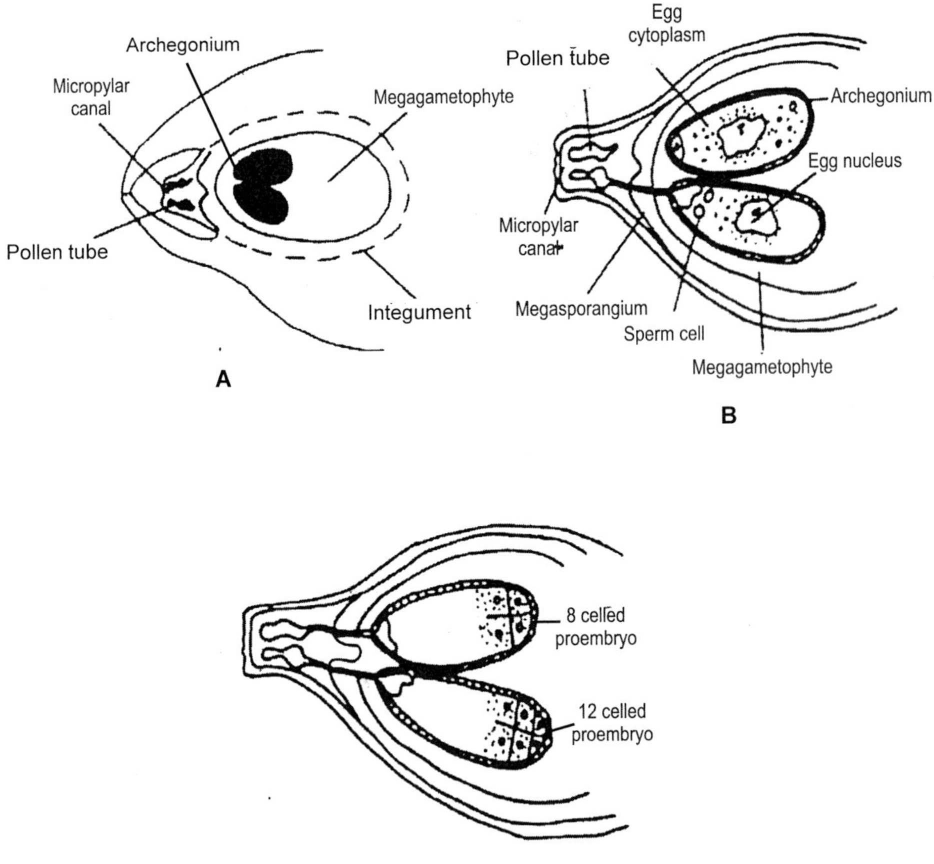

Figure 5.19 A-C: *Pinus*. L.S. Ovule - Diagrammatic sketches.

A - Formation of archegonia, post pollination stage. B - The male gametes have been delivered in egg cytoplasm. C - Proembryo formation.

- The embryo is in the form of a short straight axis.
- Its radicle is present towards the micropylar end **(Plate V.15).**
- Plumule is present towards the inner side and is surrounded by several cotyledons **(Fig.5.20; Plate V.15)**.
- The unutilized prothallial tissue forms the endosperm.
- The persistent nucellar tissue near the micropylar end forms the perisperm.
- The integument becomes hard testa.
- Some part of the ovuliferous scale fuses with the developing seed and makes a large wing for dispersal of seed.
- The axis of the female cone rapidly increases and produces gaps in ovuliferous scales.
- The cone becomes woody for the dispersal of seeds.

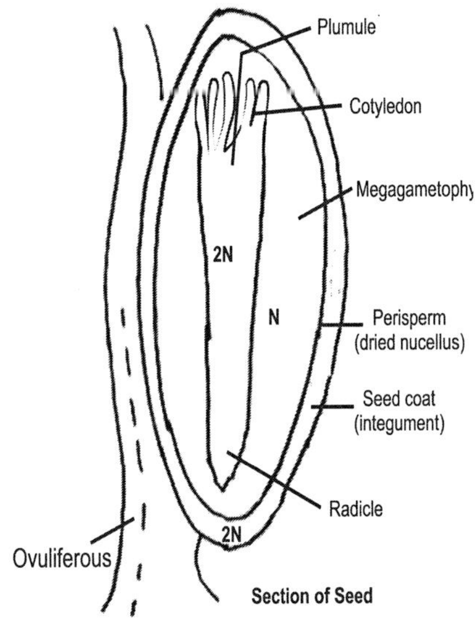

Figure 5.20: *Pinus* L.S. Seed to show polycotyledonary condition and the embryonal axis. The haploid (N) and diploid (2N) tissues belonging to three generations are marked.

Germination of Seed

- The germination of seed is epigeal.
- The radicle grows out and splits the testa at the micropylar end. It forms the primary root.
- The hypocotyl elongates to form a loop and becomes straight. It carries with it the plumule and the cotyledons.
- The testa is also carried up with the cotyledons.

Concise account of life cycle of *Pinus* is given in Figure 5.21.

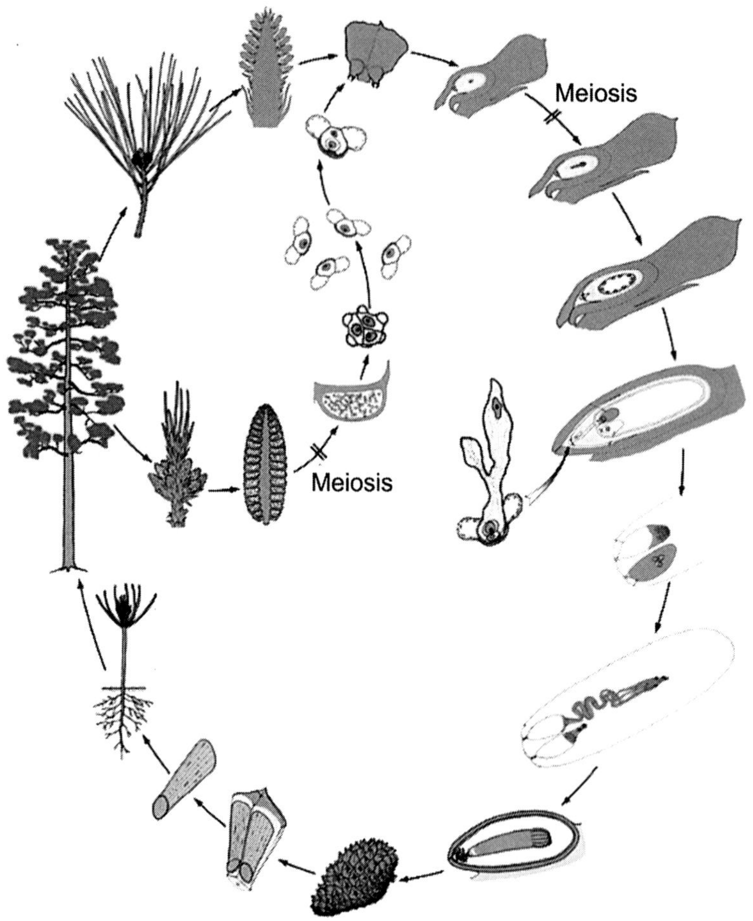

Figure 5.21: *Pinus* life cycle.

Test Your Knowledge

Q 1. Give the salient features of family Pinaceae.

Q 2. Taking example of *Pinus* explain the anatomy of conifer wood.

Q 3. What is a bract-seed-scale complex?

Q 4. Give a concise account of pollen grain development of *Pinus*.

Q 5. How is pollination and fertilization brought about in *Pinus*?

Q 6. Explain the embryogeny in *Pinus*.

Chapter 6

Gymnosperms and Human Welfare

Though gymnosperms as a group have great economical and ecological value, it is conifers which have pervaded into all spheres of human welfare. Coniferous trees and shrubs play a key role in human society and provide wealth of both wood and non-wood products. They have been revered by many cultures and have become an integral part of human folklore, mythology and religion. Some conifers have served as political symbols **(Fig. 6.1A&B)**, while others have been important in a variety of art forms.

As Ornamentals

Gymnospermous plants like *Cycas*, *Thuja*, *Araucaria* and *Juniperus* are widely used as ornamentals and planted in gardens and parks for their aesthetic appeal. In general, besides being popular ornamentals in parks, they are planted in cemeteries, and other public places, as well as around private homes. The bonsai or topiary of conifers are popular indoor plants. The traditional Christmas tree (species of conifer) of western Europe and North America is known to bring fragrance and freshness of forest into homes during the depth of the northern winters. Yews and junipers, the low-growing gymnosperms are cultivated for ground cover. Conifers can be planted as effective windbreaks. Cycads are used as garden plants in warmer latitudes, and some may even thrive indoors. Their leathery green foliage and sometimes colorful strobili are striking. *Ginkgo* is a hardy tree, and although it once approached extinction, it is now cultivated extensively and survives challenging habitats like busy streets. Some gymnosperms like Pines and Junipers are able to invade disturbed areas or abandoned agricultural land.

Figure 6.1 A&B: Gymnosperms as political symbols.

A - Coat of arms of the municipalitie of Berolle (Switzerland) with two pines depicted.
B - The coat of arms of Vermont, the official armorial bearings of the U.S. state of Vermont with pine tree.

As Food

Seeds from some gymnosperms are often used as food sources but the value for food is much less and may said to be marginal when compared to industrial uses of gymnosperms. Seeds of *Pinus* and three species of *Araucaria* are edible. *Pinus pinea* and *P. koraiensis* seeds are used for processed food. In USA and Mexico, *Pinus combroides, P. edulis, P. monophylla* are pinyon pines to be consumed. Seeds of *Pinus gerardiana* (chilgoza) are edible. Roasted pine seeds are a delicacy eaten plain or used as a garnish on bakery products.

Sago starch is extracted from the pith of sago palm stems, *Metroxylon sagu* and other cycads such as *C. revoluta*. It is a major staple food for the locals of New Guinea and the Moluccas, where it is called *saksakand sagu*. It is cooked and eaten in various forms, such as rolled into balls, mixed with boiling water to form a paste, or as a pancake. Sago looks like many other starches, and is produced commercially in form of "pearls". Sago flour from *Metroxylon* is nearly pure carbohydrate with very little protein, vitamins, or minerals.

'Kaffir bread' is a bread-like starchy food prepared from the centre of the stem of *Encephalartos*, a cycad native to Africa. For this reason, several species of *Encephalartos* are commonly referred to as bread trees, (https://en.wikipedia.org/wiki/Encephalartos - cite_note-sae1-1 bread palms or kaffir bread). The genus name is derived from the Greek words *en* (within), *kephali* (head), and *artos* (bread), referring to the use of pith to make food.

Young leaves of *Cycas* are consumed as vegetables.

Seeds of *Ginkgo* and cycads may be poisonous unless detoxified. Although *Cycas* seeds are frequently consumed as a source of carbohydrates, they contain toxic compounds cycasin (0.2–0.3 per cent) and neocycasin (methylazoxymethanol β-D-glycosides), which have adverse effects on the nervous system. The indigenous people, however know how to detoxify these for consumption.

Berries (in reality the fleshy cones) of juniper are used to flavour gin.

As Medicines

Ephedrine (alkaloid) extracted from *Ephedra* is used against asthma, cough, cold and bronchitis. Tincture of *Ephedra* is a cardiac stimulant. Young leaves of *Cycas* sp. are a source of juice, used as a cure for blood vomiting and flatulence of stomach. In traditional Chinese medicine, cycad seeds are used to treat hypertension, musculoskeletal disorders, gastrointestinal distress, cough, and amenorrhea. *Cycas* gum is used as antidote for snake bites and malignant ulcers. Two chemotherapy taxane drugs against cancer were originally developed from yew trees. Docetaxel (Taxotere) was first made from the needles of the European yew tree, *Taxus baccata* and paclitaxel (Taxol) was made from the bark of the Pacific yew tree *Taxus brevifolia* Nutt. Though both drugs are now made synthetically, the needles are still collected and used as part of the process of drug making.

As Wood

Conifers provide about 60 per cent of all wood used for industrial purposes. They enjoy technical and economic advantage over angiosperms as the gymnosperm wood is soft when compared to angiospermic plants. Gymnosperms account for approximately 75 percent of the world's timber and a large amount of the wood pulp is used to make paper. Out of this, major contributors are conifers. Most of the commercial lumber in Northern Hemisphere, the regions dominated by conifers, is derived from the trunks of conifers such as pine, douglas fir, spruce, fir, and hemlock. *Araucaria*, *Agathis australis*, and *Podocarpus* are important conifers of the Southern Hemisphere used for lumber.

The straight-grained wood of several species is light for its strength, and can be easily worked upon. In addition to its use in building construction, gymnosperm wood is used for utility poles and railroad ties. Aromatic wood of cedar is frequently used in the construction of closets or clothes chests and apparently repels cloth-eating moths. Many Cupressaceae have scented wood with less resin and is resistant to decay. In China it is used in temples and religious buildings. Other spruce wood is used to manufacture violins and similar string instruments because the wood produces a desired resonance. White pines provide higher quality, straight grain and soft wood for furniture doors and instruments like piano keys. Wood from spruce (*Picea*) is used to make log houses and church floors. The Douglas fir, *Pseudotsuga menziesii*, provides more timber than any other North American tree species and produces some of the most desirable lumber in the world. The wood is strong and relatively free of knots. Firs (*Abies*) produce light weight, soft high grade plywood. Uses of the wood include house framing, plywood production, structural beams, pulpwood, railroad ties, boxes, and crates. The wood from the redwood *Sequoia* sp. is used for furniture, fences, posts, some construction, and has various garden uses. Wood from *Agathis* gives timber known as kauri which is used for indoor construction and panelling. The plywood from *Podocarpus* is known as podo and is easy to saw and plane.

In east Asia, war victims have been buried since ancient times in coffins made from sugi (*Cryptomeria japonica*) and sanmu (*Cunninghamia lanceolata*), redwood (*Sequoia sempervirens*) and bald cypress (*Taxodium distichum*) that are equally decay and termite-resistant. Medieval archers drew long bows of the elastic yew wood (*Taxus baccata*). The eastern red cedar or pencil cedar (*Juniperus virginiana*) is used for pencil making.

As Paper Pulp

Other processed softwood products include paper and plastics derived from chemically treated wood pulp of spruces (*Picea*). Fibres of conifers make up paper pulp. In North America the white spruce, *Picea glauca*, is the main source of pulpwood used for newsprint and other paper. Papers like newsprints, writing and printing papers are being prepared from the wood pulp of *Pinus, Picea, Abies,* and *Gnetum*.

As Paints and Varnish

From conifers, commercially important resins and their distilled products are derived such as turpentine, rosin, and wood alcohol (methanol). Tannins from the bark of hemlock (*Tsuga canadensis*) and naval stores (including turpentine) from many pines are also obtained. A hardened form of resin from a kauri (*Agathis australis*), called copal, is used in the manufacture of paints and varnishes. Some resins, such as balsam (from hemlock – *Abies balsamia*) and dammar (from *Agathis*) are source of mounting media for microscope slides. Fossilized resin of conifers is a source of Amber, an anti-blood coagulant. *Cycas* gum is used as adhesive. Tannins extracted from the bark of old trees are used in leather industry. Conifer bark especially from *Araucaria, Pinus* and *Sequoia* is often the source of compounds involved in the leather tanning industry. Resin has many uses; it is used by baseball pitchers to improve their grip on the ball and by batters to improve their grip on the bat; violinists apply resin to their bows to increase friction with the strings; dancers apply resin to their shoes to improve their grip on the stage.

The Fijians used resin from an indigenous conifer as an ingredient of a glue to build their ocean going canoes. The indigenous people of some North American cultures also used conifer resins to caulk their canoes.

As Oil Used in Scents

Conifers are important sources of a wide range of non-wood products. They come from virtually every part of the tree, the foliage, bark, roots, resin, seeds and cones. The wood, foliage and resin of a number of conifers yield essential oils, which are important ingredients in perfumes, disinfectants, and cleaning products.

Commercially useful oils are derived from conifers - junipers, pines, hemlock, fir, spruces, and arborvitae. These oils serve as air fresheners, disinfectants, and scents in soaps and cosmetics. The red cedar wood oil extracted from the heart wood of *Juniperus virginiana* is used for cleaning microscopic preparations and for oil immersion lenses. Gin is flavoured from oil of *Juniperus* species. Oils obtained from *Cedrus deodara*, *Cryptomeria japonica* and *Cupressus sermperivirens* are used in preparations of perfumes.

Miscellaneous Uses

The leaves of cycads are used for preparing baskets, mats, hats and brooms, while fibres obtained from the leaves are used for stuffing pillows and making mattresses. Some gymnosperms may occasionally be used for creating artificial silk or other textiles. Conifers are frequently planted in reforestation projects. The use of pine straw and bark chips as mulches and soil amendments is used frequently. In forest, gymnosperms are important in erosion control, protection of water sheds and enhancement of aesthetic value of natural communities. Conifer forests are a source of additional non-wood products. The edible mushrooms, have the fruiting bodies of ectomycorrhizal fungi which are associated with roots of various conifers. The Japanese matsutake mushroom, which grows in forests of *Pinus densiflora* in Japan, is presently regarded as the world's most expensive food.

Certain lichens grow on stems and trunks of conifers and have been used as a traditional food, a dye source, livestock feed and for decoration. The parasitic dwarf mistletoes, which are locally important sources of livestock fodder are associated with conifers. In at least one instance, a pine feeding caterpillar is a traditional food item.

Chapter 7

Collection and Processing

Since gymnosperms are woody and several are trees, with xeric leaves, the collection and processing is different from that of other archegoniates-bryophytes or ferns, which are mostly thalloid or delicate foliose. Most gymnosperms are pressed and dried and later mounted on stiff sheets of paper or herbarium sheets. Various techniques are employed to attach the dry and fragile plant parts to sheets.

A concise account for laboratory processing of the gymnosperms is provided. However, it needs to be understood that, processing and preservation of gymnosperms requires the basic field kit for collection and common tools on field such as field note books, labels and field descriptions remain same as for delicate specimens of liverworts and mosses.

Collection of Woody Plants

In gymnosperms, like *Abies spectabilis*, the specimens need to be dipped in glycerin before pressing the leaves tend to fall. The large female cone of certain gymnosperms such as those of *Agathis* species disintegrate when mature or when dried out. To overcome this, the cone is given a coating of varnish soon after collecting, which prevents disintegration of the cone provided it is handled gently. Most of the introduced species have large female cones, *e.g.*, *Pinus*, and these may be prevented from opening on drying by varnishing also. Specimens with large woody cones should not be placed in the press with other specimens unless they are packed up till level, but it is easier to press them separately. The cones can be simply displayed in boxes after drying and processing. When specimens are collected, it is much more convenient to have a plant press out in the field and the specimens are placed in the press between sheets of paper as they are collected.

A simple press consists of two frames constructed of light wooden slats, held together by two leather straps which, when tightened, supply the pressure for pressing. The herbarium sheet on which the specimen is finally mounted measures 16½in × 10½in (Kew Standard) and thus the press is made to measure 17in × 11in. A common press has five pieces lengthwise (17in) and three pieces crosswise (11in), one at each end and one across the centre. It is provided with two leather straps that measure about four feet long and at least one inch wide and buckle holes starting at about two feet from the buckle itself.

If the plants are small then the whole plant may be collected but if it is a large plant then only a portion can be taken. It is often desirable to take a photograph of a large plant to substantiate the specimen. When a specimen is collected, a label can be attached with details of specimen number and notes against the corresponding number in field notebook. Field notes should include life form and habit, precise location, ecological data, diseases or parasites encountered on plants, and flower and fruiting characters which generally get lost in future pressing and drying. The press should be unbuckled every day and the specimens removed from folders in which they have dried, and drying papers are changed. Specimens are placed in a warm place to dry thoroughly before using again. A well prepared specimen dries out quickly but specimens can be ruined if too much heat is applied, over-dried specimens become very brittle. When all the specimens of a collection have dried they are taken from the press, arranged in numerical sequence and kept in their folders until they can be mounted. To make a specimen folder, a piece of newspaper measuring 17in × 22in is folded once to make a folder of 17in × 11in. The specimen is placed in this folder ready for pressing. As this is the approximate size of the sheet on which the specimen will eventually be mounted, the folder gives a good idea for arranging the specimen. Ends of the specimen must not protrude from the folder, if the specimen is too long it must then be bent upon itself once or more to fit within the given space. When the specimen has been mounted and the species determined, it may then be labelled. A label must contain the name of the collector and specimen number, name of the plant (when known), precise locality, and date of the collection. The value of the specimen increases if additional data is incorporated in the basic label, *e.g.*, habitat and altitude, form of the plant, notes on reproductive structures, and common names. This additional data is obtained from the field notebook.

Preservation

'Herbarium', can be a collection of dry or wet flattened specimens devoid of their colour. These specimens are prone to attack by insects or fungi, which cause putrefaction and bacteria, which cause decay. Hence, the necessity arises to preserve the specimens.

Wet Preservation

Leaves are immersed in boiling water for a few minutes and preserved in 5 per cent formalin. Colour is preserved with copper acetate and glacial acetic acid.

Dry Preservation

Cones are poisoned by using saturated solution of alcohol with mercuric chloride. Specimens are fastened to the sheet with fevicol and the label is put.

Herbarium specimens of one genus should be put in a cover. Periodical poisoning should be done. If moisture is seen even on one specimen, the entire collection should be poisoned. Naphthalene, which keeps the herbarium pests away to a certain extent can be used. They can then be preserved in the fumigation chamber.

Seed Extraction and Storage

This storage is best in racking made of light poles, which allows air to circulate around the sacks from all sides. The ease of extracting seed from cones varies according to species. Cones of noble fir break up soon after collection. Western hemlock and Western red cedar cones can be opened by spreading them in shallow boxes or trays in a well-ventilated shed. Cones of most other species are normally dried in a special seed kiln in which the temperature of the air is progressively raised as drying proceeds. After drying, the seeds are separated by shaking through coarse sieves or by rotating them in a wire-mesh drum, and then collected in a tray. Seeds with detachable wings are de-winged by gentle rubbing between cloth sheets or in a machine especially designed for the purpose. Wings, cone scales, empty seed and other impurities are separated from the seed by winnowing. The seed is then ready for testing for germination percentage, moisture content and storage. Seeds can be satisfactorily stored in this condition until the following spring, provided it is kept cool, in sealed containers. However, seeds for long term storage should be dried in warm dry air until the moisture content is between six and eight per cent. The dried seeds should then be placed in airtight containers and stored in a cold store.

Staining

Following sectioning, a temporary or permanent mount can be made in routine class work and sections can be cut with a sharp blade or a 'razor'.

The primary objective of staining a plant material is to differentiate various cell types constituting the tissue system. Each cell type has a specific physical, chemical and biological property. Based on their properties, the material takes up a particular colour. Stain also increases refractive index of the material, thus, enabling a better viewing of the material. For academic purposes, the commonly used stains are safranin and fast green. The preparation of a semi-permanent mount, requires the plant material to be subjected through a series containing alcohol and xylene. The sectioned material is initially stained. The material needs to be stained excessively in safranin as safranin is soluble in alcohol. The subsequent changes in alcohol must be carried out quickly. Various steps involved are: staining in safranin (prepared in alcohol), dehydration in graded alcohol series - alcohol 10 per cent, 20 per cent, 35 per cent, 50 per cent, 65 per cent, 75 per cent and 100 per cent alcohol. This is followed by counter staining in fast green and washing of excess stain in alcohol.

This is subsequently subjected to clearing in alcohol: xylene series – by passing through 25 per cent, 50 per cent (3:1, 1:1 and 1:3, alcohol:xylene)and then xylene 100 per cent with mounting in DPX (Distyrene, Plasticizer, and Xylene). Instruments used for changing the series must be moisture free and changes made in xylene should be done in totally moisture free atmosphere otherwise it may result in 'clouding'. The material is subjected to xylene treatment so as to increase clarity of material and also to increase refractive indices. Finally, the material is transferred to a clean slide with a drop of (DPX) and a cover slip. Such a slide can be labelled and preserved for a long time.

Need for Digital Herbaria

For more than 300 years, the pressed and dried plant specimens have been fundamental artifact in the global survey of plant diversity. The pressed plant specimens have been used as the standard for botanical preservation, storage, and comparative study (DeWolf, 1968). With changing times, climate change and human interference, biodiversity has undergone a great change. Many gymnosperm genera are left with only a few individuals and taking them away from their habitat for collection will only hasten their disappearance *e.g., Ginkgo.* Photographs which became common in late 19th century, have been little more than an ancillary part of traditional herbarium collections. They were initially accepted as a form of botanical illustrations rather than herbarium specimens and were therefore stored in botanical libraries (Simpson and Barnes, 2008). The advent of digital photography in the 1990s did not alter the principal role of photographs and they continued to be a form of illustration rather than of documentation. With increased usage of digital camera and high resolution photographs, digital herbaria are coming up in a big way. Photographic standards for type images have now been published (Häuser *et al.,* 2005). A few herbaria have made digital images of their general collections by using these digital tools to go beyond the limits of types and time. Some herbaria store both the images of dried specimens, and digital images that were taken of living plants before the photographed plant was pressed and mounted. The images are then coded and stored as a linked component to the specimen. With all the refinements and space benefits provided by digital herbaria, they are making their way into collections. The major specimens that help taxonomic studies however, are those referred from collecting, pressing, drying and then securing specimens on a stiff paper with labelling voucher.

Plate I *Cycas*

1. A crown of newly formed leaves showing circinate vernation.

2. Dichotomously branched, apogeotropic coralloid roots.

3. T.S. coralloid root showing algal zone (az) in the cortical region.

4 A&B. T.S. leaflet showing xeric characters - thick cuticle (arrows) and sunken stomata (double arrows). *B* is a higher magnification of the midrib region to show transfusion tissue (tt) and mesarch xylem.

5 . T.S. rachis with distinct 'Omega' arrangement of vascular bundles. Also seen is epidermis followed by a well differentiated hypodermis and cortex. Inset shows a few bundles enlarged.

6. Male plant with young male strobilus.

7. A portion of strobilus enlarged to show microsporophyll with sterile tip, the apophysis (ap) and the fertile portion with sori (so).

8 A&B. T.S. microsporphyll to show microsporangia on the undersurface. Mucilage cavities abound the sterile portion. In *B* the microsporangia are enlarged and are interspersed with sterile hairs (arrow) which is a fern character.

9. Female plant with a megasporophyll crown.The ovule bearing structures are loosely arranged.

10. L.S. ovule to show micropyle (my), pollen chamber (pc) and nucellar beak (nb).

11. The mature seeds are colorful often red, purple or yellow.

12 A&B. L.S. seed with massive endosperm (en), has a thick seed coat (sc) and embryonal axis (ea). In *B* photomontage shows an embryonal axis with two distinct cotyledons (co) and plumule (pl).

Embryogeny: Is the study of the formation and development of an embryo.

Endarch condition: When the primary xylem matures from center of the stem toward outside.

Eusporangiate sporangia: Having sporangia which rise from a group of epidermal cells.

Excurrent condition: Where main stem is undivided as the spruce and other conifers.

Flange: At the time of pollination, a circular rim or an umbrella-shaped structure, called "flange" develops from inner integument in the ovule of *Gnetum*.

Haplocheilic stomata: In some gymnosperms, applied to a type of stoma in which the two guard cells are derived from a single mother cell and the subsidiary cells are derived from a different initial.

Leptosporangiate sporangia: Seen in *Cycas* (of ferns) having each sporangium developing from a single cell, rather than from a group, and normally with specialized explosive spore dispersal.

Manoxylic wood: Characterised by softer secondary vascular tissues. They have large parenchyma in the form of cortex and pith; xylem tissue is less. Seen in gymnosperms like *Cycas*.

Megasporogenesis: The formation of megaspores inside the ovules of seed plants. A diploid cell in the ovule, called a megasporocyte or a megaspore mother cell, undergoes meiosis and gives rise to four haploid megaspores.

Mesarch xylem: The condition in which protoxylem in a primary xylem strand develops first in the centre of the strand and continues to develop both centrifugally and centripetally, *e.g.*, in shoots of ferns as a result, the older cells are surrounded by the younger cells. Seen in *Cycas* leaf and rachis.

Ovuliferous scale: Large woody specialized leaves that form the female cone of conifers and related trees and bear ovules, which develop into seeds. It is borne in the axils of the spirally arranged woody bracts or bract scales that constitute the cone. The ovuliferous scale is believed to be a highly specialized megasporophyll and is homologous with a carpel. It is due to the closure and opening of these scales that pollination to seed dispersal events happen.

Pollen chamber: A small chamber at the apex of the nucellus or the micropylar end in some plants (as most gymnosperms) for the reception of the pollen. Pollen may be retained here prior to fertilization.

Pollination drop (PD): It is a characteristic feature of many wind pollinated gymnosperms, is an ovular secretion that forms a landing site for pollen in many gymnosperms.

Prothallial cell: In gymnosperms and certain pteridophytes, the smaller of two cells formed by the first asymmetric division of the microspore nucleus. It also sometimes undergoes further division and represents the vegetative tissue of the gametophyte.

Rachis: The main axis of a pinnately compound leaf or of a fern is also called rachis.

Resin Canal: A canal like intercellular space in gymnosperms and some angiosperms that is lined with epithelial cells and secretes resin.

Rosette embryo: In some gymnosperms (*e.g., Pinus*), proembryo forms four distinct tiers *viz.,* embryonal tier, suspensor tier, rosette tier and open tier. Additional embryos develop from the rosette tier which generally do not survive but bring polyembryony in *Pinus*. This type of polyembryony has been termed as rosette polyembryony.

Siphonogamy: It is a condition in plants in which pollen tubes are developed for the transfer of the male gametes to the eggs. The seed plants are siphonogamous, while in the lower plants the male cells usually swim to the eggs. As a consequence, the spermatophytes were sometimes called siphonogams. see zooidogamy

Strobilus: Is a structure present on many land plant species consisting of sporangia-bearing sporophylls densely aggregated along a stem. They are many a times equated with cones but are not woody.

Sulphur shower: The winged pollen grains of *Pinus*, yellow in colour are dispersed in spring. Since these are many in number, when they are dispersed the view is a yellow dust in the air and is sulphur shower.

Syndetocheilic stomata: In some gymnosperms, applied to a type of stoma in which the two guard cells and the subsidiary cells are all derived from a single mother cell.

Transfusion tissue: It is found characteristically around the vascular bundles of gymnosperm leaves and consists of both living cells like those of parenchyma with walls that are not lignified and thin-walled but lignified tracheids with bordered pits. Seen in stems with xerophytic adaptation.

Zooidogamy: It is a type of plant reproduction in which male gametes (antherozoids) swim in a film of water to the female gametes (archegonium). Zooidogamy is found in algae, bryophytes, pteridophytes, and some gymnosperms (others use siphonogamy).

Question Bank

Q1. Fill in the Blanks

(a) The sperm cells of ________ are perhaps the largest of all known male cells in the plant kingdom.

(b) Turpentine and resin are obtained from ________ .

(c) *Gnetum* belongs to the subclass ________ .

(d) Pavement tissue is found in the nucellus of ________ .

(e) ________ is called chilgoza pine.

(f) The resin of *Abies balsamia* with refractive index of glass is ________ .

(g) The anti cancer drug Taxol is derived from ________.

(h) Is a gymnosperm without archegonium ________.

(h) Is a gymnosperm with eccentric growth rings ________ .

(i) Male cones of *Pinus* are modified ________ .

(j) Tetrasporic type of female gametophyte is found in ________ gymnosperm.

(k) Seed germination in *Cycas* is ________

Q2. Answer in detail:

(a) Why is *Cycas* called a 'living fossil'?

(b) With the help of diagram/s, describe the ovule of *Cycas* at the time of fertilization.

(c) Discuss the economic importance of *Pinus*.

(d) Discuss the affinities of *Gnetum*.

(e) With the help of well labelled outline diagrams, compare the development of male gametophyte in *Cycas, Pinus* and *Ephedra*.

(f) Give an account of life cycle of *Cycas*.

(g) Through neat well labelled diagrams, compare the ovules of *Cycas* and *Gnetum* at the time of fertilization.

(h) Compare the structures of ovules of *Pinus, Ephedra* and *Gnetum* at the time of fertilization.

(i) Assign reasons for placing *Ephedra* and *Gnetum* in separate orders.

(j) *Ephedra* is considered advanced over *Cycas*.

(k) Enumerate the diagnostic characteristics of Gymnosperms. How do they resemble and differ from Pteridophytes?

(l) 'Absence of secondary veins in leaves of gymnosperms is compensated by the presence of transfusion tissue'. Elaborate with reference to gymnosperms studied by you.

Q3. Draw well labelled diagrams of the following:

(a) T.S. *Cycas* leaflet.

(b) L.S. *Ephedra* ovule at the time of fertilization.

(c) T.S. *Pinus* needle.

(d) L.S. *Gnetum* male strobilus.

(e) L.S. *Ephedra* male strobilus.

(f) L.S. *Ephedra* seed (mark three different generations).

(g) T.S. *Cycas* Coralloid root.

Q3. Distinguish between:

(a) Centripetal and centrifugal xylem.

(b) Primary and secondary transfusion tissue.

(c) Manoxylic and Pycnoxylic wood.

(d) Dwarf shoot and long shoot.

Q4. Write short notes on:

(a) Coralloid roots.

(b) Bract-Seed-Scale-Complex.

(c) Pollen of *Pinus.*

(d) Polyembryony in *Pinus.*

(e) Anomalous secondary growth in *Gnetum ula.*

(f) Male strobilus of *Ephedra.*

(g) Structure of microsporophyll and microsporangia in *Cycas*

(h) Post pollination changes in *Pinus.*

Suggested Readings

Begoviæ Bego, B.M. 2011. Nature's Miracle *Ginkgo biloba*. Book 1 (Vols. 1–2) pp. 60–61.

Biswas, C. and Johri, B.M. 1997. The Gymnosperms. Springer Science and Business Media, Science 494 pages.

Bold, H.C., Alexopoulo C.J. and Delevoryas, T. 1987. Morphology of Plants and Fungi, 5th ed. Harper and Row, New York, USA.

Bolinder, K., Humphreys, A.M., Ehrlen, J., Alexandersson, R., Ickert-Bond, S.M., and Catarina, R. 2016. From near extinction to diversification by means of a shift in pollination mechanism in the gymnosperm relict *Ephedra* (Ephedraceae, Gnetales). *Bot J of the Linnean Soc* 180: 461–477.

Bond, W.J. 1989. The tortoise and the hare - Ecology of angiosperm dominance and gymnosperm persistence. *Biol J of Linnean Soc* 36: 227-249.

Brenner, E.D., Stevenson, D.W., and Twigg, R.W. 2003. Cycads: Evolutionary innovations and the role of plant-derived neurotoxins. *Trends in Plant Sci.* 9: 446-52.

Brenner, E.D., Stevenson, D.W., McCombie, R.W., Katari, M.S., Rudd, S.A., Mayer, K.F., Palenchar, P.M., Runko, S.J., Twigg, R.W., Dai, G., Martienssen, R.A., Benfey, P.N., and Coruzzi, G.M. 2003. Expressed sequence tag analysis in *Cycas*, the most primitive living seed plant. *Genome Biol.* 4(12): R78.

Brodribb, T.J. and Feild, T.S. 2010. Leaf hydraulic evolution led a surge in leaf photosynthetic capacity during early angiosperm diversification. *Ecol Lett*13: 175-183.

Brummitt, N.A., Bachman, S.P., Griffiths-Lee, J., Lutz, M., Moat, J.F., and Farjon, A. 2015. Green plants in the red: A baseline global assessment for the IUCN sampled red list index for plants. *PLOS ONE* 10 (8): e0135152.

Brundrett, M.C. 2009. Mycorrhizal associations and other means of nutrition of vascular plants: Understanding the global diversity of host plants by resolving conflicting information and developing reliable means of diagnosis. *Plant Soil* 320(1): 37-77.

Byng, J.W. 2015. The Gymnosperms Hand Book: Practical Guide to Extant Families and Genera of World. Plant Gateway Ltd UK.

Carmichael, J.S. and Friedman, W.E. 1995. Double fertilization in *Gnetum gnemon*: The relationship between the cell cycle and sexual reproduction. *Plant Cell.* 7 (12): 1975-1988.

Chamberlain, C.J. 1935. Gymnosperms: Structure and Evolution. The University of Chicago Press, Chicago.

Chase, M.W. and Reveal, J.L. 2009. A phylogenetic classification of the land plants to accompany APG III. *Bot J Linnean Soc* 161: 122-127.

Christenhusz, M., Reveal, J.L., Farjon, A., Gardner, M.F., Mill, R.R. and Chase, M.W. 2011. A new classification and linear sequence of extant gymnosperms. *Phytotaxa* 19(1):55-70.

Doyle, J.A. 1996. Seed plant phylogeny and the relationships of Gnetales. *Int J Plant Sci* 157 (Supplement 6): S3 – S39.

Doyle, J.A., and Donoghue, M. J. 1986. Seed plant phylogeny and the origin of angiosperms: An experimental cladistic approach. *Bot Rev* 52: 321 – 431.

Endress, P. K. 1996. Structure and function of female and bisexual organ complexes in Gnetales. *Int J Plant Sci* 157: S113-S125.

Friedman, W.E. 1990. Sexual reproduction in *Ephedra nevadensis* (Ephedraceae): Further evidence of double fertilization in a non-flowering seed plant. *Amer J Bot* 77: 1582-1598.

Gilbert, N. 2010. Threats to the world's plants assessed. *Nature* 10.1038/news.499.

Gong, Y.B., Yang, M., Vamosi, J.C., Yang, H.M., Mu, W.X., Li, J.K., and Wan, T. 2015. Wind or insect pollination? Ambophily in a subtropical gymnosperm *Gnetum parvifolium* (Gnetales). *Plant Species Bio* 31: 272-279.

Govil, C.M. 2007. Gymnosperms Extinct and Extant. Krishna Prakashan Media (P) Ltd Delhi.

Ickert-Bond, S.M., Skvarla, J.J. and Chissoe, W.F. 2003. Pollen dimorphism in *Ephedra* L. (Ephedraceae). *Rev of Palaeobot and Palynol* 124: 325-334.

Isaac, N. J. B., Turvey, S. T., Collen, B., Waterman, C. and Baillie, J. E. M. 2007. Mammals on the EDGE: Conservation priorities based on threat and phylogeny. *PLOS ONE* 2(3): e296.

Jones, D.L. 1993. Cycads of the World: Ancient Plants in Today's Landscape. Smithsonian Institution Press, Washington, DC.

Kozlowski, G., Stoffel, M., Bétrisey, S., Cardinaux, L. and Mota, M. 2015. Hydrophobia of gymnosperms: Myth or reality? A global analysis. *Ecohydrology* 8: 105-112.

Kozlowski, T.T. (ed). 1972. Seed Biology: Importance, Development, and Germination. Volume 1, Academic Press Inc, New York.

Lee, C. 1995. "Fertilization in *Ginkgo biloba*," *Bot Gaz* 117: 79-100.

Leslie, A., Beaulieu, J.M. and Mathews, S. 2017. Variation in seed size is structured by dispersal syndrome and cone morphology in conifers and other nonflowering seed plants. *New Phytol* 216(2): 429-437.

Lill, B.S. and Sweet, G.B. 1977. Pollination in *Pinus radiata. N Z J Forestry Sci* 7(1): 21-34.

Liston, A., Gernandt, D.S., Vining, T.F., Campbell, C.S. and Pin˜ero, D. 2003. Molecular phylogeny of Pinaceae and *Pinus. Acta Hortic* 615: 107–114.

Meyer, S.E. 2008. *Ephedra* L.: ephedra or Mormon-tea.

Bonner, F.T., Karrfatt, R.P., (eds). The Woody Plant Seed Manual. Agriculture Handbook No. 727. Washington, D.C. US Department of Agriculture, Forest Service. pp. 492-494.

Mundry, M. and Stützel, T. 2004. Morphogenesis of the reproductive shoots of *Welwitschia mirabilis* and *Ephedra distachya* (Gnetales), and its evolutionary implications. *Org Divers & Evol* 4: 91–108.

Nagalingum, N.S., Marshall, C.R., Quental, T.B., Rai, H.S., Little, D.P. and Mathews, S. 2011. Recent synchronous radiation of a living fossil. *Science* 334 (6057): 76-79.

Prantl, K.A.E. 2010. Lehrbuch der Botanik. Kessinger Publishing, LLC.

Richardson, D.M. (ed.). 1998. Ecology and Biogeography of *Pinus.* Cambridge University Press, New York, USA.

Russell, S.D. and Christian, D. 1992 (ed). Sexual Reproduction in Flowering Plants, Volume 140 (International Review of Cytology. A Survey of Cell Biology). Academic Press.

Ryberg, P.E., Rothwell, G.W., Stockey, R.A.,Hilton, J.,Mapes, G., and Riding, J.B. 2012. Reconsidering relationships among stem and crown group Pinaceae: Oldest record of the genus *Pinus* from the early cretaceous of Yorkshire, UK. *Int J Plant Sci* 173: 917- 932.

Rydin, C., Khodabandeh, A. and Endress, P.K. 2010. The female reproductive unit of *Ephedra* (Gnetales): Comparative morphology and evolutionary perspectives. *Bot J of Linnean Soc* 163:387–430.

Rydin, C., Pedersen, K.R., Crane, P.R., and Friis, E.M. 2006. Former diversity of *Ephedra* (Gnetales): Evidence from Early Cretaceous seeds from Portugal and North America. *Ann Bot* 98: 123–140.

Sarvas, R. 1962: Investigations on the flowering and seed crop of *Pinus sylvestris. Metsantutkimuslaitoksen julkaisuja* 53: 1-198.

Smertenko, A.P., Bozhkov, V., Filonova, L.H., von Arnold, S. and Hussey, P.J. 2003. Re-organization of the cytoskeleton during developmental programmed cell death in *Picea abies* embryos. *Plant J* 33: 813–824.

Stevens, P.F. 2001. Angiosperm Phylogeny Website.

Stevenson, D. 1990. A formal classification of the ectant cycads. *Brittonia* 44(2): 220-223.

Stevenson, W. 1993. Ephedraceae. In: Flora of North America Editorial Committee (eds.): *Flora of North America North of Mexico Vol.* 2. Oxford University Press pp. 428-434.

Takaso, T. and Bouman, F. 1986. Ovule and seed ontogeny in *Gnetum gnemon* L. The botanical magazine = Shokubutsu-gaku-zasshi 99: 241–266.

Takaso, T., Kimoto, Y., Owens, J.N., Kono, M., and Mimura, T. 2012. Secretions from the female gametophyte and their role in spermatozoid induction in *Cycas revoluta. Plant Rep* 26: 17–23. doi: 10.1007/s00497-012-0204-5.

Thureborn, O. and Rydin, C. 2015. Phylogeny of *Ephedra* revisited s. Botany. Edmonton, Alberta, Canada, Abstract ID: 1010.

William, C.G., Auckland, L.D., Reynolds, M.M., and Leach, K.A. 2003. Overdominant lethals as part of the conifer embryo lethal system. *Heredity* 91(6): 584-592.

Winter, K.U., Becker, A., Münster, T., Kim, J.T., Saedler, H., and Theissen, G. 1999. MADS-box genes reveal that gnetophytes are more closely related to conifers than to flowering plants. *Proc Natl Acad of Sci* USA 96: 7342-7347.

Yong, Y. and Wang, Qi. 2013.The earliest fleshy cone of *Ephedra* from the Early cretaceous Yixian formation of northeast China. *PLoS ONE*. 8: e53652.

Zhou, Z., and Zheng, S. 2003. The missing link in *Ginkgo* evolution. *Nature* 423(6942): 821-2.

Internet Sources

blogs.discovermagazine.com//living-fossil-cycad-plants-are-actually-evolutions-com.

Ginkgo - Ginkgo biloba - Details - Encyclopedia of Life eol.org/pages/1156278/details).

http://faculty.ycp.edu/~kkleiner/fieldnaturalhistory/labs/Lab1_Class_conifers/Gymnosperm_naked_seed.jpg.

http://www.conifers.org/zz/Cycadales.php.

https://artdesign.rit.edu/faculty-staff/37/faculty/1022.

https://wikivisually.com/wiki/*Cycas_sphaerica.*

https://www.edgeofexistence.org.

https://www.researchgate.net/publication/228691404_A_new_classification_and_linear_sequence_of_extant_gymnosperms.

https://www.researchgate.net/publication/260760282_Somatic_embryogenesis_Life_and_death_processes_during_apical-basal_patterning.

https://www.triarchincorporated.com/product/ephedra-stemprepared-microscope-slide-10-3/.

www.conifers.org/ep/Ephedraceae.php.

http://e-pao.net/epSubPageExtractor.asp?src=education.Science_and_Technology.Biodiversity_of_gymnosperms.

http://www.mobot.org/MOBOT/Research/APweb/orders/gnetales.

http://www.plantapalm.com/vce/taxonomy/classification.htm(6-Jan-1999).

Index

A

Adiantum 50

Alveoli 8, 120

Antherophore 68, 86, 87

Archegonium 4, 39, 57, 90-92, 122-125, 152, 154

Anemophilous 8, 9

Anatropous ovule 149

B

Bars of Sanio 51, 111

Bordered pits 25, 51, 64, 77, 77, 84, 101, 107, 109, 111, 152, 154

Bracts 14, 67, 86, 87, 94, 95, 98, 104, 153

Bryophytes 4, 18, 134, 154

C

Caffeine 97

Cedar 5, 131-133, 136

Centrifugal xylem 29, 30, 31, 157

Chichi roots 50

Classification *v*, 15, 17, 19, 101, 103, 159, 161, 162

Coenomegaspore 71, 72, 80

Collar 7, 54, 67, 68, 69, 80, 141, 143

Conifers 1, 2, 5-10, 20, 47, 62, 77, 81, 99-103, 129-133

Coralloid root 20, 23, 24, 28, 152, 156, 157

Corallorrhiza 28

Cyas circinalis 38

Cycadidae 2, 19, 20, 21

Cycads 1-3, 5, 6, 8-11, 17, 18, 20-24, 37, 38, 41, 44, 47, 62, 100, 129-133, 152, 158, 160, 161

Cypress knees 12, 13

Cytoplasmic dense zone 96

D

Digital herbaria 137

Dispersules 4

Double fertilization 62, 63, 73, 96, 97, 98

Dry preservation 136

E

Ectomycorrhizal mycelia 112

Embryogeny 14, 42, 59, 74, 93, 94, 124

Encephalartos laurentianus 21

Ephedrales 80

Ephedra nevadensis 85, 159

Ephedrine 95, 97, 98, 131
Extra-ovular capture 39

F

Female cone 117
Female gametophyte 6, 36, 37, 40, 41, 56, 57, 63, 71, 72, 73, 90, 91, 94, 119, 121, 152, 155, 161
Female strobilus 35, 54, 69, 87
Fir 5
Flange 70, 71, 75, 81, 113, 153, 143, 149
Food 130

G

Ginkgo adiantoides 46
Ginkgo biloba 2, 45-48, 56
Ginkgoidae 2, 19, 45, 46, 47
Ginkgos 2, 46, 100
Gnetidae 2, 18, 19, 61, 62
Gnetophytes 2, 6, 7, 10, 14, 17, 46, 56, 61, 62, 63, 81, 161
Gnetum gnemon 62, 65, 73, 80
Gymnosperm database 3, 10, 17, 20
Gymnosperm evolution 10
Gynodioecious 7

H

Haplocheilic 85, 153
Human welfare 129
Hydrophobes 12

I

Inner phellogen 28

J

Juniper *x*, 5, 101, 129, 131, 132

L

Lepidozamia hopei 21
Living fossils 2, 21, 22

M

Macrozamia dyeri 21
Male gametophyte 4, 5, 36, 38, 55, 71, 77, 90, 119, 120, 156
Male strobilus 32, 33, 47, 53, 68, 86, 87, 92, 143
Maule reaction 79
Medicines 131
Megasporogenesis 71, 89
Mesarch xylem 25, 30, 31, 138, 153
Metasequoia glyptostroboides 2
Microsporangium 5, 8, 33, 34, 53, 54, 69, 87, 115, 116
Microsporophyll 33, 34, 53, 115, 116
Miscellaneous uses 133
Manoxylic wood 25, 44, 153
Mormon tea 6, 81, 95, 97

N

Naked seeded plants 5, 6, 16
Node 20, 50, 63, 80, 82-87
Non-wood products 129, 133

O

Ovulate strobilus 54, 87
Ovuliferous scale 117, 121, 127

P

Parasitaxus usta 11
Perianth 4, 63, 68, 69, 70, 79, 86, 87, 89, 143, 146
Pinidae 2, 19, 99
Pinus gerardiana 103, 130
Pinus lambertiana 11
Pinus longaeva 7, 101
Pinus monophylla 104
Pinus roxburghii 103
Pollen capture mechanisms 39

Pollination 37, 57

Pollination drop *vi*, 5, 8, 9, 14, 39, 56, 57, 58, 73, 89, 92, 100, 120, 122, 141, 148, 153

Polyembryony 9, 14, 44, 74, 76, 93, 97, 123-125, 154, 157

Polyembryos 42

Pycnoxylic wood 107, 109, 157

Preservation 135

Proembryo 42, 59, 74, 75, 93, 94, 97, 102, 124, 154, 126

Prothallial cell 9, 14, 36, 39, 40, 55, 57, 71, 79, 90, 98, 119, 153

R

Rachis 23, 30, 31, 138, 153, 154

Redwood 6, 11, 132

Resin 5, 62, 84, 103, 104, 106, 107, 109, 111, 112, 113, 118, 121, 131-133, 154, 155

Reproductive structures 1, 8, 9, 14, 33, 53, 62, 67, 86, 115, 116, 135, 152

Rosette tier 125, 154

S

Sago starch 25, 130

Scents 133

Secondary growth 25, 65, 84, 106

Secondary tissues 25, 26, 65, 110

Seed-scale complex 102, 121, 128

Sequoiadendron giganteum 6, 11

Sequoia sempervirens 11, 132

Siphonogamy 4, 41, 42, 154

Spermatozoids 5, 8, 20, 21, 36, 38, 40-42, 44, 57, 58, 58, 62, 100, 152

Spongy parenchyma 29, 83, 113

Sporophyte 23, 24, 48, 64, 82, 104, 105

Spur shoots 48, 49, 50, 53

Strobili 1, 8, 14, 37, 54, 69, 76, 77, 80, 82, 85-88, 92, 93, 94, 95, 98, 129

Strobilus *vi*, 8, 21, 32, 33, 35, 44, 47, 53, 4, 66, 68, 69, 86, 87, 92, 94, 95, 152, 154, 156, 157

Sulphur shower 120, 148, 149, 154

Suspensor cells 74, 75, 93, 94, 125, 126

T

Tannins 37, 97, 132

Transfusion tissue 28-30, 52, 113, 138, 154, 156, 157

Transmitting tissue 91

Taxodium distichum 12, 13, 132

Taxus baccata 7, 131, 132

Thuja plicata 7

U

V

Vegetative reproduction 11, 32, 52, 85, 114

Vessels 7, 13, 14, 31, 62-66, 77, 79, 80, 84

W

Wet preservation 135

Wollemia nobilis 2

Wood 131

Z

Zooidogamy 8, 41, 42, 154